MATRIX ANALYSIS OF CIRCUITS USING MATLAB®

JAMES G. GOTTLING

Ohio State University

PRENTICE HALL, Englewood Cliffs, NJ 07632

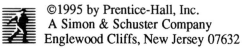

©1995 by Prentice-Hall, Inc.
A Simon & Schuster Company
Englewood Cliffs, New Jersey 07632

Matlab is a registered trademark of The MathWorks, Inc.

The MathWorks, Inc.
Cochituate Place
24 Prime Park Way
Natick, Massachusetts 01760

Printed in the United States of America

10 9 8 7 6 5 4 3 2 1

ISBN 0-13-127044-3

Prentice-Hall International (UK) Limited, London
Prentice-Hall of Australia Pty. Limited, Sydney
Prentice-Hall Canada Inc., Toronto
Prentice-Hall Hispanoamericana, S.A., Mexico
Prentice-Hall of India Private Limited, New Delhi
Prentice-Hall of Japan, Inc., Tokyo
Simon & Schuster Asia Pte. Ltd., Singapore
Editora Prentice-Hall do Brasil, Ltda., Rio de Janeiro

Table of Contents

Preface

This book shows how to solve linear circuit problems using MATLAB®. An acronym for Mathematics Laboratory, MATLAB is an easy-to-use, command-line driven computer program. Versions of this program exist for MS-DOS, Macintosh, and UNIX computers.

Prior to the early 1970's, engineering students, like knights errant, went to their classes with slide rules strapped to their belts. The electronic calculator changed that scene virtually overnight; the shirt pocket became the repository of calculating capability. Engineering faculties then faced the momentous decision whether to allow or disallow use of calculators on midterms and exams. In retrospect, that question is trivial. Everyone has and uses calculators. The important issues are how to set up problems and how to interpret their results. As universities acquired powerful mainframe computers, learning computer programming became a part of a young engineer's training. The invention of the microprocessor and development of inexpensive personal computers is once again about to change the engineering education environment. Electronic calculators, like buggy whips and slide rules, will become relics. Personal pocket computers exist now and will become more powerful. These hardware developments have led to software developments as well. We have seen the development of FORTRAN, BASIC, PLI, APL, PASCAL, C, MODULA, and C++. However, engineering computations often are done using simulation or problem-solving programs rather than general-purpose programming languages. In electrical engineering at this time the major simulation or problem-solving programs are SPICE for electronic circuit simulation and MATLAB for matrix computations.

Students will learn circuit theory from the viewpoint of analysis, simulation, and computation. Learning simulation techniques is essential for effective preparation for electronics courses, while developing computational skills is specially important for control and communication theory courses. SPICE is the dominant circuit simulation program. In my opinion, MATLAB is the best matrix-solving program.

This book has four chapters. The first chapter describes matrix representation of linear equations, matrix manipulation, and numerical solution methods for linear equations. If you are already familiar with these topics, go directly to Chapter 2. The second chapter is a MATLAB tutorial that focuses on MATLAB's ability to do tasks that are useful in circuit analysis. If you understand how to use MATLAB, continue to the third chapter. The third chapter shows how to write DC and AC circuit equations directly by inspection of a circuit diagram using nodal analysis, mesh analysis or modified nodal analysis (MNA), even for a circuit that has controlled sources. You will find some new MATLAB *Circuit Toolbox* functions in this chapter. These new functions simplify entry of circuit elements into node, mesh, or MNA equations. Using these new functions often eliminates the necessity of writing the circuit equations in the first place. The final chapter shows how to construct Bode plots and to obtain transient solutions for circuits using MATLAB. Appendix A lists the new MATLAB circuit functions.

All of the example M-files and the new circuit M-file functions are on a computer diskette that accompanies this manual. This diskette is a 3.5-inch MS-DOS format high-density diskette. To use these files on an MS-DOS computer, create subdirectories within your MATLAB subdirectory on your hard disk, naming these CK_TOOL and CK_EX. Copy all of the files in the CK_TOOL subdirectory on the diskette into your new CK_TOOL hard-disk subdirectory and all of the files in the CK_EX diskette subdirectory into your new hard-disk CK_EX subdirectory. From MATLAB, use the `path` command to be sure that MATLAB can find these files.

To use these files on a Macintosh, copy the MS-DOS diskette onto your Macintosh hard disk using the Apple File Exchange program. Depending on the version of Macintosh System software that you are using, you can find the *Apple File Exchange* program in the *Apple File Exchange Folder* within the *Apple Utilities* folder on either the *Tidbits* or *Utilities* system disk. Run the Apple File Exchange program and copy the subdirectories CK_EX and CK_TOOL from the diskette into the folder on your hard disk that contains your version of MATLAB. Before you copy these files, select the **MS-DOS to Mac** menu and remove the check mark from the **MacWrite to DCA-RFT...** item and place a check mark before the **Text translation...** item. Leave all items in the **Text translation** dialog as you find them. If you are using Version 4.X of MATLAB, you can read the examples or run them from MATLAB's **File** menu. If you are using The Student Edition of MATLAB, you can open the example files from the **File** menu but can not immediately use **Run Script...** on the **M-File** menu to run the examples. Open the example that you want to run. Then, click on **Save and Go** on the **File** menu to run the example. To save the example so that you can run it later, with the example file open select **Save as...** from the **File** menu. Delete the suffix .M from the file name; e.g., change EX_2_2.M to EX2_2 in the **Save as...** dialog box. Now you can run file EX_2_2 directly using **Run Script...** from the **M-File** menu. If you have any problem running any of the examples that use circuit toolbox functions, be sure that MATLAB has the folders CK_EX and CK_TOOL in the path (See the **Set Path** item on the **M-File** menu in Version 3.5 or on the **File** menu in Version 4.X).

I am grateful to my students at The Ohio State University, who listened patiently to my lectures during the development of the node and mesh circuit inspection methods that appear in Chapter 3. Also, I wish to thank my wife, Dr. Sue Gottling, who patiently gave me free time on weekends to write this book. Her compensation for my preoccupation was to shop for clothes for our four grandchildren. Finally, I dedicate this manual to the well-clothed grandchildren, Sarah Croft Cottrill, Cyrus James Politi, Katherine Swan Cottrill, and Nikkita Helena Gottling.

MATRIX ANALYSIS OF CIRCUITS USING MATLAB®

Chapter 1

Determinants and Matrices

This chapter describes properties of determinants and matrices that you need to understand to use MATLAB effectively. Determinants and matrices are powerful tools for analysis of linear equations. Since analysis of linear circuits involves linear equations, knowing how to use determinants and matrices effectively is essential for solving linear circuits.

1.1 Determinant Definition and Properties

In this section we identify a determinant using three defining properties. These defining properties lead to three additional important properties. In addition, this section defines a determinant's minors and cofactors. These are useful in the numerical evaluation of a determinant.

1.1.1 Definition

A determinant $A = A(a_{ij})$, where $i, j = 1\ldots n$, written in the form

$$A = A(a_{ij}) = \begin{vmatrix} a_{11} & a_{12} & \cdots & a_{1n} \\ a_{21} & a_{22} & \cdots & a_{2n} \\ \vdots & \vdots & \ddots & \vdots \\ a_{n1} & a_{n2} & \cdots & a_{nn} \end{vmatrix} \qquad (1.1)$$

has the following three defining properties[1]

1. The value of determinant A remains the same if the sum of the elements of a row (column) and the corresponding elements of any other row (column) replace the elements of that row (column).

2. The value of determinant A changes by a factor k if all elements of any row (column) change by the factor k.

3. The value of determinant A equals unity if all elements on the principal diagonal (a_{kk}, where $k = 1\ldots n$) are unity while the off-diagonal elements are zero.

These three fundamental properties lead to the following additional properties

4. The value of determinant A remains the same if the sum of the elements of a row (column) and the product of an arbitrary factor times the corresponding elements of any other row (column) replace the elements of any row (column).

5. The algebraic sign of determinant A reverses if the elements of any two rows (columns) interchange.

---

[1] E.A. Guillemin, *The Mathematics of Circuit Analysis*, John Wiley & Sons, Inc., New York, 1949.

6. The value of the determinant A is zero if all the elements of any row (column) are zero, or if the corresponding elements of any two rows (columns) are the same or have a common ratio.

1.1.2 Minors and Cofactors

A first-order minor M_{ij} results from determinant A on deletion of the i–th row and j-th column. Refer to the element a_{ij} as the complement of the minor M_{ij}. The cofactor A_{ij} has the value

$$A_{ij} = (-1)^{i+j} M_{ij} \qquad (1.2)$$

1.1.3 Laplace Evaluation of Determinant

The Laplace row expansion of determinant A is

$$A = \sum_{j=1}^{n} a_{ij} A_{ij} \qquad (1.3)$$

where row index i assumes any value from 1 to n. Alternately, the Laplace column expansion is

$$A = \sum_{i=1}^{n} a_{ij} A_{ij} \qquad (1.4)$$

where column index j assumes any value from 1 to n.

1.2 Matrix Solution of Linear Equations

A set of n linear equations relating the response values y_i to the excitation values x_i is

$$
\begin{aligned}
a_{11}y_1 + a_{12}y_2 + \ldots + a_{1n}y_n &= x_1 \\
a_{21}y_1 + a_{22}y_2 + \ldots + a_{2n}y_n &= x_2 \\
\vdots \qquad \vdots \qquad \quad \vdots \quad \vdots \\
a_{n1}y_1 + a_{n2}y_2 + \ldots + a_{nn}y_n &= x_n
\end{aligned}
\qquad (1.5)
$$

Defining matrices $\mathcal{A}$, $\mathcal{Y}$, and $\mathcal{X}$

$$
\mathcal{A} =
\begin{bmatrix}
a_{11} & a_{12} & \ldots & a_{1n} \\
a_{21} & a_{22} & \ldots & a_{2n} \\
\ldots & \ldots & \ldots & \ldots \\
a_{n1} & a_{n2} & \ldots & a_{nn}
\end{bmatrix}
\qquad (1.6)
$$

$$\mathcal{Y} = \begin{bmatrix} y_1 & y_2 & \ldots & y_n \end{bmatrix}^t \qquad (1.7)$$

$$\mathcal{X} = \begin{bmatrix} x_1 & x_2 & \ldots & x_n \end{bmatrix}^t \qquad (1.8)$$

where the superscript "t" denotes *transpose*, which changes the row matrix into a column matrix. In this matrix notation, the linear equations become

$$\mathcal{A} * \mathcal{Y} = \mathcal{X} \tag{1.9}$$

For i = 1... n, the matrix product $\mathcal{A}*\mathcal{Y}$ gives

$$\sum_{j=1}^{n} a_{ij} y_j = x_i \tag{1.10}$$

1.2.1 Cramer's Rule

The Cramer's rule solution of a set of linear equations comes from the relations

$$a_{1i}A_{1k} + a_{2i}A_{2k} + \ldots + a_{ni}A_{nk} = \begin{cases} A & i = k \\ 0 & i \neq k \end{cases} \tag{1.11}$$

$$a_{i1}A_{k1} + a_{i2}A_{k2} + \ldots + a_{in}A_{kn} = \begin{cases} A & i = k \\ 0 & i \neq k \end{cases} \tag{1.12}$$

which are evident; because for i = k they represent Laplace expansions for columns and rows respectively, and for i ≠ k they represent Laplace expansions for a determinant possessing identical rows and columns respectively.

Multiplying Eqs. 1.5 successively by the cofactors A_{1k}, A_{2k}, ... , A_{kk}, ... , A_{nk}, adding, and collecting coefficients of like terms

$$
\begin{aligned}
&(a_{11}A_{1k} + a_{21}A_{2k} + \ldots + a_{n1}A_{nk})y_1 \\
&+ (a_{12}A_{1k} + a_{22}A_{2k} + \ldots + a_{n2}A_{nk})y_2 \\
&+ \ldots \\
&+ (a_{1k}A_{1k} + a_{2k}A_{2k} + \ldots + a_{nk}A_{nk})y_k \\
&+ \ldots \\
&+ (a_{1n}A_{1k} + a_{2n}A_{2k} + \ldots + a_{nn}A_{nk})y_n = A_{1k}x_1 + A_{2k}x_2 + \ldots + A_{nk}x_n
\end{aligned} \tag{1.13}
$$

for k = 1, ... , n. These reduce to

$$y_k = \frac{A_{1k}}{A} x_1 + \frac{A_{2k}}{A} x_2 + \ldots + \frac{A_{nk}}{A} x_n \tag{1.14}$$

Equation 1.14 can be written using the notation

$$\mathcal{Y} = \mathcal{A}^{-1} * \mathcal{X} = \frac{\text{adj}(\mathcal{A})}{\det(\mathcal{A})} * \mathcal{X} \tag{1.15}$$

Here det($\mathcal{A}$) is the determinant of matrix $\mathcal{A}$, and the matrix inverse $\mathcal{A}^{-1}$ satisfies the equation

$$\mathcal{A}^{-1} * \mathcal{A} = I = \mathcal{A} * \mathcal{A}^{-1} \tag{1.16}$$

The identity matrix I is square, has the same size as $\mathcal{A}$, consists of ones on it's diagonal, and has zeros elsewhere. The adjoint of matrix $\mathcal{A}$ [adj($\mathcal{A}$)] is

$$\mathrm{adj}(\mathcal{A}) = \begin{bmatrix} A_{11} & A_{21} & \cdots & A_{n1} \\ A_{12} & A_{22} & \cdots & A_{n2} \\ \cdots & \cdots & \cdots & \cdots \\ A_{1n} & A_{2n} & \cdots & A_{nn} \end{bmatrix}$$

(1.17)

To form the adjoint of matrix $\mathcal{A}$, first construct a matrix from the cofactors of the elements a_{ij}, then take the transpose of this matrix. The transpose results from interchanging elements across the diagonal (A_{ij} and A_{ji}).

1.2.2 Gaussian Elimination

Evaluation of the solution $\mathcal{Y}$ from the excitation $\mathcal{X}$ using Cramer's rule for matrix inversion is numerically inefficient. Gaussian elimination is a procedure that gives a numerically efficient solution of a set of equations. This procedure begins by appending the x_i values as an $(n+1)$-th column to the $n \times n$ array of coefficients a_{ij}. Next, without changing the solution values, linear operations convert the array into triangular form with values of one on the diagonal. This conversion begins with diagonal index k set to one and proceeds, incrementing k by one at each step until k equals n. Each step has two parts. The first part normalizes the k-th equation's diagonal coefficient a_{kk} to one by dividing all element values on and to the right of the diagonal by a_{kk}. The second part changes the k-th column coefficients of equations k+1 to n to zero by replacing all a_{ij} by $a_{ij} - (a_{ik} \times a_{kj})$ for i equal to k+1 to n and j equal to k to n+1.

The final processing step solves the set of diagonal equations by replacing the $(n+1)$-th column of elements by the solution vector $\mathcal{Y}$. This process, called *back substitution*, finds the solution values by solving from the $(n-1)$-th equation down to the first equation using variables already in place. Formally, this process is

$$a_{i,n+1} = a_{i,n+1} - \sum_{j=i+1}^{n} a_{ij} \times a_{j,n+1}$$

(1.18)

Sometimes during triangularization a zero value occurs at the current diagonal position. If this happens then, to avoid division by zero, the process has to interchange the current row coefficients with those of any subsequent row having a nonzero coefficient in the k-th column. The best numerical accuracy occurs if this interchange occurs with the subsequent row having the largest-valued numerical coefficient in the k-th column. Even when the current diagonal position is nonzero, using this interchange strategy improves the solution's numerical accuracy. This interchange step is known as *partial pivoting*.

Example 1.1

Using Gaussian elimination, determine the solution of the matrix equation

$$\mathcal{A} * \mathcal{Y} = \begin{bmatrix} a_{11} & a_{12} & a_{13} \\ a_{21} & a_{22} & a_{23} \\ a_{31} & a_{32} & a_{33} \end{bmatrix} * \begin{bmatrix} y_1 \\ y_2 \\ y_3 \end{bmatrix} = \begin{bmatrix} 4 & -2 & 0 \\ -2 & 4 & -2 \\ 0 & -2 & 4 \end{bmatrix} * \begin{bmatrix} y_1 \\ y_2 \\ y_3 \end{bmatrix} = \begin{bmatrix} 1 \\ 0 \\ 0 \end{bmatrix} = \begin{bmatrix} x_1 \\ x_2 \\ x_3 \end{bmatrix} = \mathcal{X}$$

Solution

Augment the 3×3 matrix A into a 3×4 matrix using the X matrix as a fourth column of coefficients. Then

$$\begin{bmatrix} 4 & -2 & 0 & \vdots & 1 \\ -2 & 4 & -2 & \vdots & 0 \\ 0 & -2 & 4 & \vdots & 0 \end{bmatrix} \overset{1}{\Rightarrow} \begin{bmatrix} 1 & -1/2 & 0 & \vdots & 1/4 \\ -2 & 4 & -2 & \vdots & 0 \\ 0 & -2 & 4 & \vdots & 0 \end{bmatrix} \overset{2}{\Rightarrow} \begin{bmatrix} 1 & -1/2 & 0 & \vdots & 1/4 \\ 0 & 3 & -2 & \vdots & 1/2 \\ 0 & -2 & 4 & \vdots & 0 \end{bmatrix}$$

$$\overset{3}{\Rightarrow} \begin{bmatrix} 1 & -1/2 & 0 & \vdots & 1/4 \\ 0 & 1 & -2/3 & \vdots & 1/6 \\ 0 & -2 & 4 & \vdots & 0 \end{bmatrix} \overset{4}{\Rightarrow} \begin{bmatrix} 1 & -1/2 & 0 & \vdots & 1/4 \\ 0 & 1 & -2/3 & \vdots & 1/6 \\ 0 & 0 & 8/3 & \vdots & 1/3 \end{bmatrix}$$

$$\overset{5}{\Rightarrow} \begin{bmatrix} 1 & -1/2 & 0 & \vdots & 1/4 \\ 0 & 1 & -2/3 & \vdots & 1/6 \\ 0 & 0 & 1 & \vdots & 1/8 \end{bmatrix} \overset{6}{\Rightarrow} \begin{bmatrix} 1 & -1/2 & 0 & \vdots & 1/4 \\ 0 & 1 & -2/3 & \vdots & 1/4 \\ 0 & 0 & 1 & \vdots & 1/8 \end{bmatrix}$$

$$\overset{7}{\Rightarrow} \begin{bmatrix} 1 & -1/2 & 0 & \vdots & 3/8 \\ 0 & 1 & -2/3 & \vdots & 1/4 \\ 0 & 0 & 1 & \vdots & 1/8 \end{bmatrix}$$

The first step normalizes the first row by dividing each coefficient to the right of a_{11} by the value of a_{11}. The second step reduces the column of coefficients below a_{11} to zero by replacing the coefficients in each row below a_{11} with

$$a_{ij} \Rightarrow a_{ij} - a_{i1} \times a_{1j} \qquad i = 2,3; \, j = 1,\ldots,4$$

The third step normalizes the second row, dividing each coefficient in this row by the current value of a_{22}. The fourth step reduces the column of coefficients below a_{22} to zero by replacing the elements in each row below a_{22} with

$$a_{ij} \Rightarrow a_{ij} - a_{i2} \times a_{2j} \qquad i = 3; \, j = 2,\ldots,4$$

The fifth step normalizes the last row, giving

$$\begin{bmatrix} 1 & -1/2 & 0 \\ 0 & 1 & -2/3 \\ 0 & 0 & 1 \end{bmatrix} \begin{bmatrix} y_1 \\ y_2 \\ y_3 \end{bmatrix} = \begin{bmatrix} 1/4 \\ 1/6 \\ 1/8 \end{bmatrix}$$

The third row of coefficients in this matrix equation gives

$$y_3 = 1/8$$

Starting back substitution, step six uses the second row of coefficients to obtain

$$y_2 = \frac{1}{6} - \left(-\frac{2}{3}\right)\left(\frac{1}{8}\right) = \frac{1}{4}$$

Finally, from the first row of coefficients, step seven gives

$$y_1 = \frac{1}{4} - \left(-\frac{1}{2}\right)\left(\frac{1}{4}\right) - 0\left(\frac{1}{8}\right) = \frac{3}{8}$$

```
/* Gauss solution
      Define variables n, np1 = n+1, and nm1 = n-1 elsewhere.
      Uses zero-order rows and columns.
*/

for (k = 0; k < n; k++){
      /* Normalize k-th row */
      for (j = k + 1; j < np1; j++){
            a[k][j] /= a[k][k];
      }
      /* Set k-th column to zero for rows k+1 to n */
      for (i = k + 1; i < n; i++) {
            for (j = k + 1; j < np2; j++) {
                  a[i][j] -= a[i][k] * a[k][j];
            }
      }
}

/* Back substitution */
for (i = nm1; i >= 0; i--) {
      for (j = i + 1; j < np1; j++) {
            a[i][np1] -= a[i][j] * a[j][np1];
      }
}
```

Figure 1.1 Gaussian Elimination Method.

Figure 1.1 shows a C-language version of the Gaussian procedure. This implementation of the Gaussian procedure avoids division of a_{kk} by itself in the normalization step by starting the column index j at k+1 instead of k. Also, setting all of the column elements below the diagonal element to zero is not done. These simplifications are possible because the back substitution procedure assumes that each diagonal element is one and that the coefficient values below each diagonal element are all zero. This program does not do partial pivoting, so it can not solve a set of equations that has a zero anywhere on the diagonal.

1.2.3 $\mathcal{LU}$ Factorization and Crout Method

The Crout[2] method, using $\mathcal{LU}$ factorization, is another numerically efficient procedure to determine the solution of a set of linear equations. This procedure factors a matrix $\mathcal{A}$, having the form

$$\mathcal{A} = \begin{bmatrix} a_{11} & a_{12} & \dots & a_{1n} \\ a_{21} & a_{22} & \dots & a_{2n} \\ \vdots & \vdots & \ddots & \vdots \\ a_{n1} & a_{n2} & \dots & a_{nn} \end{bmatrix}$$

(1.19)

into the product of two matrices

$$\mathcal{A} = \mathcal{L} * \mathcal{U}$$

(1.20)

---

[2] P. D. Crout, , "A short method for evaluating determinants and solving systems of linear equations with real or complex coefficients," AIEE Transactions, Vol. 60, pp1235-1240, 1941.

The lower-triangular matrix $\mathcal{L}$ is

$$\mathcal{L} = \begin{bmatrix} l_{11} & 0 & \cdots & 0 \\ l_{21} & l_{22} & \cdots & 0 \\ \vdots & \vdots & \ddots & \vdots \\ l_{n1} & l_{n2} & \cdots & l_{nn} \end{bmatrix} \qquad (1.21)$$

and the upper-triangular matrix $\mathcal{U}$ is

$$\mathcal{U} = \begin{bmatrix} 1 & u_{12} & \cdots & u_{1n} \\ 0 & 1 & \cdots & u_{2n} \\ \vdots & \vdots & \ddots & \vdots \\ 0 & 0 & \cdots & 1 \end{bmatrix} \qquad (1.22)$$

Then

$$\mathcal{A} = \mathcal{L}*\mathcal{U} = \begin{bmatrix} l_{11} & 0 & \cdots & 0 & 0 \\ l_{21} & l_{22} & \cdots & 0 & 0 \\ \vdots & \vdots & \ddots & \vdots & \vdots \\ l_{n-1,1} & l_{n-1,2} & \cdots & l_{n-1,n-1} & 0 \\ l_{n1} & l_{n2} & \cdots & l_{n,n-1} & l_n \end{bmatrix} \begin{bmatrix} 1 & u_{12} & \cdots & u_{1,n-1} & u_{1n} \\ 0 & 1 & \cdots & u_{2,n-1} & u_{2n} \\ \vdots & \vdots & \ddots & \vdots & \vdots \\ 0 & 0 & \cdots & 1 & u_{n-1,n} \\ 0 & 0 & \cdots & 0 & 1 \end{bmatrix} \qquad (1.23)$$

Taking the matrix product shows that the first column of elements of matrix $\mathcal{A}$ are

$$\begin{array}{ll} a_{11} = l_{11} & l_{11} = a_{11} \\ a_{21} = l_{21} & l_{21} = a_{21} \\ \vdots & \vdots \\ a_{n1} = l_{n1} & l_{n1} = a_{n1} \end{array} \qquad (1.24)$$

The first row of elements to the right of the first diagonal position are

$$\begin{array}{ll} a_{12} = l_{11}u_{12} & u_{12} = a_{12}/l_{11} \\ a_{13} = l_{11}u_{13} & u_{13} = a_{13}/l_{11} \\ \vdots & \vdots \\ a_{1n} = l_{11}u_{1n} & u_{1n} = a_{1n}/l_{11} \end{array} \qquad (1.25)$$

The second column of elements from the second diagonal position downwards are

$$\begin{array}{ll} a_{22} = l_{21}u_{12} + l_{22} & l_{22} = a_{22} - l_{21}u_{12} \\ a_{32} = l_{31}u_{12} + l_{32} & l_{32} = a_{32} - l_{31}u_{12} \\ \vdots & \vdots \\ a_{n2} = l_{n1}u_{12} + l_{n2} & l_{n2} = a_{n2} - l_{n1}u_{12} \end{array} \qquad (1.26)$$

The second row of elements to the right of the second diagonal position are

$$a_{23} = l_{21}u_{13} + l_{22}u_{23} \qquad u_{23} = (a_{23} - l_{21}u_{13})/l_{22}$$

$$a_{24} = l_{21}u_{14} + l_{22}u_{24} \qquad u_{24} = (a_{24} - l_{21}u_{14})/l_{22}$$

$$\vdots \qquad\qquad\qquad\qquad \vdots \qquad\qquad\qquad\qquad (1.27)$$

$$a_{2n} = l_{21}u_{1n} + l_{22}u_{2n} \qquad u_{2n} = (a_{2n} - l_{21}u_{1n})/l_{22}$$

The third column of elements from the third diagonal position downwards are

$$a_{33} = l_{31}u_{13} + l_{32}u_{23} + l_{33} \qquad l_{33} = a_{33} - l_{31}u_{13} - l_{32}u_{23}$$

$$a_{43} = l_{41}u_{13} + l_{42}u_{23} + l_{43} \qquad l_{43} = a_{43} - l_{41}u_{13} - l_{42}u_{23}$$

$$\vdots \qquad\qquad\qquad\qquad \vdots \qquad\qquad\qquad\qquad (1.28)$$

$$a_{n3} = l_{n1}u_{13} + l_{n2}u_{23} + l_{n3} \qquad l_{n3} = a_{n3} - l_{n1}u_{13} - l_{n2}u_{23}$$

The third row of elements to the right of the third diagonal position are

$$a_{34} = l_{31}u_{14} + l_{32}u_{24} + l_{33}u_{34} \qquad u_{34} = (a_{34} - l_{31}u_{14} - l_{32}u_{24})/l_{33}$$

$$a_{35} = l_{31}u_{15} + l_{32}u_{25} + l_{33}u_{35} \qquad u_{35} = (a_{35} - l_{31}u_{15} - l_{32}u_{25})/l_{33}$$

$$\vdots \qquad\qquad\qquad\qquad \vdots \qquad\qquad\qquad\qquad (1.29)$$

$$a_{3n} = l_{31}u_{1n} + l_{32}u_{2n} + l_{33}u_{3n} \qquad u_{3n} = (a_{3n} - l_{31}u_{1n} - l_{32}u_{2n})/l_{33}$$

A pattern emerges. As diagonal position k runs from 1 to the number of rows and columns n, form the elements l_{ik} of the $\mathcal{L}$ matrix in the column from the diagonal position downwards using

$$l_{ik} = a_{ik} - \sum_{j=1}^{k-1} l_{ij}u_{jk} \qquad\qquad i = k, k+1, \ldots, n \qquad\qquad (1.30)$$

and form the elements u_{kj} of the $\mathcal{U}$ matrix in the row to the right of the k-th diagonal position from

$$u_{kj} = \left(a_{kj} - \sum_{i=1}^{k-1} l_{ki}u_{ij} \right)/l_{kk} \qquad j = k+1, k+2, \ldots n \qquad\qquad (1.31)$$

As the process proceeds, the elements that a calculation step needs already exist above and to the left of the current row and column. Therefore the procedure can place the elements of $\mathcal{L}$ and $\mathcal{U}$ directly into the starting matrix $\mathcal{A}$, replacing the original values of the elements a_{ij}. The elements l_{ij} go into $\mathcal{A}$ from the diagonal downwards, and the elements u_{ij} replace the $\mathcal{A}$-matrix elements to the right of the diagonal. To solve

$$\mathcal{A} * \mathcal{Y} = \mathcal{L} * \mathcal{U} * \mathcal{Y} = \mathcal{L} * \mathcal{Z} = \mathcal{X} \qquad\qquad (1.32)$$

where

$$\mathcal{U} * \mathcal{Y} = \begin{bmatrix} 1 & u_{12} & \cdots & u_{1n} \\ 0 & 1 & \cdots & u_{2n} \\ \vdots & \vdots & \ddots & \vdots \\ 0 & 0 & \cdots & 1 \end{bmatrix} * \begin{bmatrix} y_1 \\ y_2 \\ \vdots \\ y_n \end{bmatrix} = \mathcal{Z} = \begin{bmatrix} z_1 \\ z_2 \\ \vdots \\ z_n \end{bmatrix} \qquad\qquad (1.33)$$

and

$$\mathcal{L} * \mathcal{Z} = \begin{bmatrix} l_{11} & 0 & \cdots & 0 \\ l_{21} & l_{22} & \cdots & 0 \\ \vdots & \vdots & \ddots & \vdots \\ l_{n1} & l_{n2} & \cdots & l_{nn} \end{bmatrix} * \begin{bmatrix} z_1 \\ z_2 \\ \vdots \\ z_n \end{bmatrix} = \mathcal{X} = \begin{bmatrix} x_1 \\ x_2 \\ \vdots \\ x_n \end{bmatrix} \qquad (1.34)$$

To find the components of matrix $\mathcal{Z}$, work successively from the first row forwards to the last in Eq. 1.34 to obtain

$$l_{11}z_1 = x_1 \qquad\qquad z_1 = x_1 / l_{11}$$
$$l_{21}z_1 + l_{22}z_2 = x_2 \qquad\qquad z_2 = (x_2 - l_{21}z_1)/l_{22}$$
$$\vdots \qquad\qquad\qquad \vdots \qquad\qquad\qquad (1.35)$$
$$l_{n1}z_1 + l_{n2}z_2 \ldots + l_{nn}z_n = x_n \qquad z_n = (x_n - l_{n1}z_1 - l_{n2}z_2 - \ldots - l_{n,n-1}z_{n-1})/l_{nn}$$

This *forward substitution* sequence of calculations generalizes to

$$z_k = \left(x_k - \sum_{j=1}^{k-1} l_{kj}z_j \right) / l_{kk} \qquad k = 1, 2, \ldots, n \qquad (1.36)$$

All elements z_j have their final values when calculating any z_k, so the elements of $\mathcal{Z}$ can replace the elements of $\mathcal{X}$. Since the process that generates $\mathcal{Z}$ is the same as the row processing steps of the $\mathcal{L}\mathcal{U}$ factorization, the $\mathcal{A}$ matrix can be augmented, placing $\mathcal{X}$ into an additional (n+1)-th row. Then, forward substitution converts the $\mathcal{X}$ vector to the $\mathcal{Z}$ vector at the same time that row processing occurs.

To obtain $\mathcal{Y}$ from $\mathcal{Z}$, work from the last row to the first in Eq. 1.33 to obtain

$$y_n = z_n \qquad\qquad y_n = z_n$$
$$y_{n-1} + u_{n-1,n}z_n = z_{n-1} \qquad y_{n-1} = z_{n-1} - u_{n-1,n}z_n$$
$$\vdots \qquad\qquad\qquad \vdots \qquad\qquad\qquad (1.37)$$
$$y_1 + u_{12}z_2 + \ldots + u_{1n}z_n = z_1 \qquad y_1 = z_1 - u_{12}z_2 - \ldots - u_{1n}z_n$$

The generalization of this sequence is

$$y_i = z_i - \sum_{j=i+1}^{n} u_{ij}z_j \qquad i = n-1, n-2, \ldots, 1 \qquad (1.38)$$

In this *back-substitution* step, all z_j elements have their final values when calculating each y_i, so the y_i values can replace the corresponding z_i elements. This in-place formation of the $\mathcal{L}$ and $\mathcal{U}$ factors, replacement of $\mathcal{X}$ with $\mathcal{Z}$, and then $\mathcal{Z}$ with $\mathcal{Y}$ is the Crout method. Figure 1.2 shows a C-language version of the Crout procedure. In this program, matrix $\mathcal{A}$ has n rows and n+1 columns. Enumeration of the rows and columns begins with 0. Column n, the (n+1)-th column in the matrix, contains the excitation vector. The C language `for` loop does the test condition before executing any statement within the loop. If the test fails immediately, then no statement in the loop executes. When a summation above has no range, then it does not exist, so the C-language `for loop` correctly takes no action for these cases.

```
/* Crout solution
      Define variables n, nm2 = n-2 elsewhere. Uses zero-
      order rows and columns.
*/

for (k = 0; k < n; k++){
      /* Do columns */
      for (i = k; i < n; i++){
            for (j = 0; j < k; j++){
                  a[i][k] -= a[i][j] * a[j][k];
            }
      }

      /* Do rows and forward elimination */
      for (j = k + 1; j <= n; j++) {
            for (i = 0; i < k; i++) {
                  a[k][j] -= a[k][i] * a[i][j];
            }
            a[k][j] /= a[k][k];
      }
}

/* Do back substitution */
for (i = nm2; i >= 0; i--) {
      for (j = i + 1; j < n; j++) {
            a[i][n] -= a[i][j] * a[j][n];
      }
}
```

Figure 1.2 $\mathcal{LU}$ Factorization Method.

In some cases the Crout procedure fails even though the equations have a solution, because an element with a zero value occurs at the diagonal position. Partial pivoting again avoids this failure and also improves the numerical accuracy of the solution. Partial pivoting examines the elements in the column below the current diagonal position to see if any have larger magnitudes than the current diagonal element. If any do, then the partial pivoting process interchanges the current row and the row having the largest element magnitude. This row interchange does not change the solution or the order of the components in the solution vector $\mathcal{Y}$.

Example 1.2

Use $\mathcal{LU}$ factorization to determine the solution of the matrix equation

$$\mathcal{A} * \mathcal{Y} = \begin{bmatrix} a_{11} & a_{12} & a_{13} \\ a_{21} & a_{22} & a_{23} \\ a_{31} & a_{32} & a_{33} \end{bmatrix} * \begin{bmatrix} y_1 \\ y_2 \\ y_3 \end{bmatrix} = \begin{bmatrix} 4 & -2 & 0 \\ -2 & 4 & -2 \\ 0 & -2 & 4 \end{bmatrix} * \begin{bmatrix} y_1 \\ y_2 \\ y_3 \end{bmatrix} = \begin{bmatrix} 1 \\ 0 \\ 0 \end{bmatrix} = \begin{bmatrix} x_1 \\ x_2 \\ x_3 \end{bmatrix} = X$$

Solution

Augment the 3×3 matrix $\mathcal{A}$ into a 3×4 matrix using the $\mathcal{X}$ matrix as a fourth column. Then

$$\begin{bmatrix} 4 & -2 & 0 & \vdots & 1 \\ -2 & 4 & -2 & \vdots & 0 \\ 0 & -2 & 4 & \vdots & 0 \end{bmatrix} \overset{1}{\Rightarrow} \begin{bmatrix} 4 & -1/2 & 0 & \vdots & 1/4 \\ -2 & 3 & -2/3 & \vdots & 1/6 \\ 0 & -2 & 8/3 & \vdots & 1/8 \end{bmatrix} \overset{2}{\Rightarrow} \begin{bmatrix} 4 & -1/2 & 0 & \vdots & 3/8 \\ -2 & 3 & -2/3 & \vdots & 1/4 \\ 0 & -2 & 8/3 & \vdots & 1/8 \end{bmatrix}$$

where

First Row :
$$a_{12} = -2/4 = -1/2$$
$$a_{13} = 0/3 = 0$$
$$a_{14} = 1/4$$

Second Column :
$$a_{22} = 4 - (-2)(-1/2) = 3$$
$$a_{32} = -2 - (0)(-1/2) = -2$$

Second Row :
$$a_{23} = \left[-2 - (-2)(0) \right]/3 = -2/3$$
$$a_{24} = \left[0 - (-2)(1/4) \right]/3 = 1/6$$

Third Column :
$$a_{33} = 4 - (0)(0) - (-2)(-2/3) = 8/3$$

Third Row :
$$a_{34} = \left[0 - (0)(1/4) - (-2)(1/6) \right]/(8/3) = 1/8$$

Back Substitution :
$$a_{24} = 1/6 - (-2/3)(1/8) = 1/4$$
$$a_{14} = 1/4 - (-1/2)(1/4) - (0)(1/8) = 3/8$$

The solution now exists in the fourth column of the 3×4 matrix of values. Notice that each calculation of a new a_{ij} starts with the original a_{ij} coefficient, in place in the matrix, and computes the new value from coefficients to the left and above the current position.

Another way to factor a matrix $\mathcal{A}$ into a product of lower- and upper-triangular factors places ones on the lower-triangular matrix diagonal, instead of on the upper-triangular matrix diagonal. In this form

$$\mathcal{A} = \mathcal{L}' * \mathcal{U}' = \begin{bmatrix} 1 & 0 & \cdots & 0 \\ l'_{21} & 1 & \cdots & 0 \\ \vdots & \vdots & \ddots & \vdots \\ l'_{n1} & l'_{n2} & \cdots & 1 \end{bmatrix} * \begin{bmatrix} u'_{11} & u'_{12} & \cdots & u'_{1n} \\ 0 & u'_{22} & \cdots & u'_{2n} \\ \vdots & \vdots & \ddots & \vdots \\ 0 & 0 & \cdots & u'_{nn} \end{bmatrix} \qquad (1.39)$$

The two forms are equivalent, since

$$\mathcal{L}*\mathcal{U} = \begin{bmatrix} l_{11} & 0 & \cdots & 0 \\ l_{21} & l_{22} & \cdots & 0 \\ \vdots & \vdots & \ddots & \vdots \\ l_{n1} & l_{n2} & \cdots & l_{nn} \end{bmatrix} * \begin{bmatrix} 1 & u_{12} & \cdots & u_{1n} \\ 0 & 1 & \cdots & u_{2n} \\ \vdots & \vdots & \ddots & \vdots \\ 0 & 0 & \cdots & 1 \end{bmatrix}$$

$$= \begin{bmatrix} l_{11}/l_{11} & 0 & \cdots & 0 \\ l_{21}/l_{11} & l_{22}/l_{22} & \cdots & 0 \\ \vdots & \vdots & \ddots & \vdots \\ l_{n1}/l_{11} & l_{n2}/l_{22} & \cdots & l_{nn}/l_{nn} \end{bmatrix} * \begin{bmatrix} l_{11} & 0 & \cdots & 0 \\ 0 & l_{22} & \cdots & 0 \\ \vdots & \vdots & \ddots & \vdots \\ 0 & 0 & \cdots & l_{nn} \end{bmatrix} * \begin{bmatrix} 1 & u_{12} & \cdots & u_{1n} \\ 0 & 1 & \cdots & u_{2n} \\ \vdots & \vdots & \ddots & \vdots \\ 0 & 0 & \cdots & 1 \end{bmatrix}$$

$$= \begin{bmatrix} 1 & 0 & \cdots & 0 \\ l_{21}/l_{11} & 1 & \cdots & 0 \\ \vdots & \vdots & \ddots & \vdots \\ l_{n1}/l_{11} & l_{n2}/l_{22} & \cdots & 1 \end{bmatrix} * \begin{bmatrix} l_{11} & l_{11}u_{12} & \cdots & l_{11}u_{1n} \\ 0 & l_{22} & \cdots & l_{22}u_{2n} \\ \vdots & \vdots & \ddots & \vdots \\ 0 & 0 & \cdots & l_{nn} \end{bmatrix}$$

$$= \begin{bmatrix} 1 & 0 & \cdots & 0 \\ l'_{21} & 1 & \cdots & 0 \\ \vdots & \vdots & \ddots & \vdots \\ l'_{n1} & l'_{n2} & \cdots & 1 \end{bmatrix} * \begin{bmatrix} u'_{11} & u'_{12} & \cdots & u'_{1n} \\ 0 & u'_{22} & \cdots & u'_{2n} \\ \vdots & \vdots & \ddots & \vdots \\ 0 & 0 & \cdots & u'_{nn} \end{bmatrix} = \mathcal{L}'*\mathcal{U}'$$

(1.40)

MATLAB uses this form of $\mathcal{LU}$ factorization.

1.3 Matrix   Algebra

This section describes the algebraic properties of matrices. These include negation, addition, and subtraction and their associative and commutative behavior. You will learn how to multiply and factor matrices, to partition matrices, and to do left- and right-matrix division.

1.3.1 Arithmetic

You can negate a matrix or add and subtract two matrices.

<u>Negation</u>

Because each equation in a set of equations can be multiplied by -1 without changing the solution, you can negate a matrix by changing the sign of all elements of the matrix, giving

$$-\mathcal{A} = -\begin{bmatrix} a_{11} & a_{12} & \cdots & a_{1n} \\ a_{21} & a_{22} & \cdots & a_{2n} \\ \vdots & \vdots & \ddots & \vdots \\ a_{n1} & a_{n2} & \cdots & a_{nn} \end{bmatrix} = \begin{bmatrix} -a_{11} & -a_{12} & \cdots & -a_{1n} \\ -a_{21} & -a_{22} & \cdots & -a_{2n} \\ \vdots & \vdots & \ddots & \vdots \\ -a_{n1} & -a_{n2} & \cdots & -a_{nn} \end{bmatrix}$$

(1.41)

Addition

Addition of two matrices requires that the matrices have the same number of rows and columns. This statement is the *conformability* rule of addition. In matrix addition, corresponding elements add, so

$$\mathcal{A} + \mathcal{B} = \begin{bmatrix} a_{11} & a_{12} & \cdots & a_{1m} \\ a_{21} & a_{22} & \cdots & a_{2m} \\ \vdots & \vdots & \ddots & \vdots \\ a_{n1} & a_{n2} & \cdots & a_{nm} \end{bmatrix} + \begin{bmatrix} b_{11} & b_{12} & \cdots & b_{1m} \\ b_{21} & b_{22} & \cdots & b_{2m} \\ \vdots & \vdots & \ddots & \vdots \\ b_{n1} & b_{n2} & \cdots & b_{nm} \end{bmatrix}$$

$$= \begin{bmatrix} a_{11} + b_{11} & a_{12} + b_{12} & \cdots & a_{1m} + b_{1m} \\ a_{21} + b_{21} & a_{22} + b_{22} & \cdots & a_{2m} + b_{2m} \\ \vdots & \vdots & \ddots & \vdots \\ a_{n1} + b_{n1} & a_{n2} + b_{n2} & \cdots & a_{nm} + b_{nm} \end{bmatrix}$$

$$(1.42)$$

Subtraction

Subtraction follows from negation and addition, so

$$\mathcal{A} - \mathcal{B} = \mathcal{A} + (-\mathcal{B}) = \begin{bmatrix} a_{11} - b_{11} & a_{12} - b_{12} & \cdots & a_{1m} - b_{1m} \\ a_{21} - b_{21} & a_{22} - b_{22} & \cdots & a_{2m} - b_{2m} \\ \vdots & \vdots & \ddots & \vdots \\ a_{n1} - b_{n1} & a_{n2} - b_{n2} & \cdots & a_{nm} - b_{nm} \end{bmatrix} \qquad (1.43)$$

Commutation

Matrix addition commutes, so

$$\mathcal{A} + \mathcal{B} = \mathcal{B} + \mathcal{A} \qquad (1.44)$$

Association

Matrix addition or subtraction associates, so

$$\mathcal{A} \pm (\mathcal{B} \pm C) = (\mathcal{A} \pm \mathcal{B}) \pm C = \mathcal{A} \pm \mathcal{B} \pm C \qquad (1.45)$$

1.3.2 Multiplication and Factoring

The conformability rule for matrix multiplication requires that the number of columns of the first matrix be the same as the number of rows of the second matrix. The usual definition of the matrix product

$$C = \mathcal{A} * \mathcal{B} \qquad (1.46)$$

defines each element of matrix C as the sum of the products of successive row elements of the first matrix $\mathcal{A}$ with corresponding column elements of the second matrix $\mathcal{B}$, which has the mathematical form

$$c_{ij} = a_{i1} b_{1j} + a_{i2} b_{2j} + \ldots a_{i,ca} b_{ca,j} = \sum_{k=1}^{ca=rb} a_{ik} b_{kj} \qquad (1.47)$$

where ca is the number of columns of $\mathcal{A}$ and rb is the number of rows of $\mathcal{B}$. The number of rows rc of matrix C equals the number of rows ra of $\mathcal{A}$, and the number of columns cc of matrix C equals the number of columns cb of $\mathcal{B}$.

Commutation
Except for special cases, matrix multiplication does not commute, so

$$\mathcal{A}*\mathcal{B} \neq \mathcal{B}*\mathcal{A} \tag{1.48}$$

Association
Matrix multiplication associates, so

$$\mathcal{A}*(\mathcal{B}*C) = (\mathcal{A}*\mathcal{B})*C = \mathcal{A}*\mathcal{B}*C \tag{1.49}$$

The product has ra rows and cc columns, ca equals rb, and cb equals rc.

Distribution
Multiplication distributes over addition or subtraction, so

$$\mathcal{A}*(\mathcal{B}\pm C) = \mathcal{A}*\mathcal{B} \pm \mathcal{A}*C \tag{1.50}$$

and

$$(\mathcal{B}\pm C)*\mathcal{A} = \mathcal{B}*\mathcal{A} \pm C*\mathcal{A} \tag{1.51}$$

The usual priority rules apply, so

$$\mathcal{A}*\mathcal{B}+C = (\mathcal{A}*\mathcal{B})+C \tag{1.52}$$

1.3.3 Partitioning
A matrix may be partitioned into submatrices. For example

$$\mathcal{A} = \begin{bmatrix} a_{11} & a_{12} & a_{13} & a_{14} & a_{15} \\ a_{21} & a_{22} & a_{23} & a_{24} & a_{25} \\ a_{31} & a_{32} & a_{33} & a_{34} & a_{35} \\ a_{41} & a_{42} & a_{43} & a_{44} & a_{45} \\ a_{51} & a_{52} & a_{53} & a_{54} & a_{55} \end{bmatrix} = \begin{bmatrix} \mathcal{A}_{11} & \mathcal{A}_{12} & \mathcal{A}_{13} \\ \mathcal{A}_{21} & \mathcal{A}_{22} & \mathcal{A}_{23} \\ \mathcal{A}_{31} & \mathcal{A}_{32} & \mathcal{A}_{33} \end{bmatrix} \tag{1.53}$$

where, for example, matrix $\mathcal{A}_{12}$ is

$$\mathcal{A}_{12} = \begin{bmatrix} a_{13} \end{bmatrix} = a_{13} \tag{1.54}$$

A matrix or a matrix partition can have no rows or no columns. In either event, the matrix is a *null* matrix. If two partitioned matrices are to be added the number of rows and columns of each corresponding partition must be the same. To multiply two partitioned matrices, the number of columns of each successive partition of the first matrix must equal the number of rows in each successive partition of the second matrix. Of course, the number of column partitions in the first matrix has to match the number of row partitions of the second matrix.

1.3.4 Matrix Division

Although matrix division often remains undefined, MATLAB defines both left- and right-division operators. The left-division operator (\) means

$$Y = A \setminus X \quad \Rightarrow \quad Y = A^{-1} * X \tag{1.55}$$

and the right-division operator (/) signifies

$$Y = X / A \quad \Rightarrow \quad Y = X * A^{-1} \tag{1.56}$$

These operators have the same priority as multiplication, with the usual left-to-right order of operations of the same priority, so

$$A + B * C \setminus D = A + \big((B * C) \setminus C \big) \tag{1.57}$$

Of course, the matrices in each operation have to be conformable.

Chapter 2

Using MATLAB

This Chapter explains how to use The Student Edition of MATLAB,[1] MATLAB 3.5, or MATLAB 4.X[2] to solve problems in linear circuit theory. The Student Edition of MATLAB, available for MS-DOS and Macintosh computers, is virtually the same as the professional version 3.5, except that vectors and matrices can not exceed 1024 elements. If available, The Student Edition of MATLAB uses a math coprocessor. The professional version operates on workstations, mainframe computers, and supercomputers in addition to Macintosh and MS-Windows personal computers. MATLAB is a matrix-processing computer program capable of working with complex numbers, so is appropriate for solving circuit problems where the circuit is describable by a set of linear equations where the coefficients may be complex numbers. The intent in this manual is to give you enough information to solve linear circuit problems using MATLAB.

2.1 Running MATLAB

This Section tells you how to install MATLAB, how to open the program, And gives general information about the nature of the program.

2.1.1 Installing

Follow the instructions that come with your disks to install MATLAB on your computer.

2.1.2 The Command Window

As shown in Fig. 2.1, MATLAB begins with a Command window that is blank except for the symbol ». This symbol is MATLAB's prompt, which shows you that MATLAB is waiting for your input. Try typing 2+2 to see the following

```
»2+2
ans = 4
```

Try some other arithmetic operators

```
»5-3
ans = 2
»2*3
ans = 6
»3/5
ans = 0.6000
»3^8
ans = 6561
»5-3
ans = 2
```

---

[1] The MathWorks, Inc., *The Student Edition of MATLAB*, Englewood Cliffs, New Jersey: Prentice Hall, 1992.
[2] MATLAB® is a registered trademark of The MathWorks, Inc., Cochituate Place, 24 Prime Park Way, Natick, Massachusetts 01760-1500.

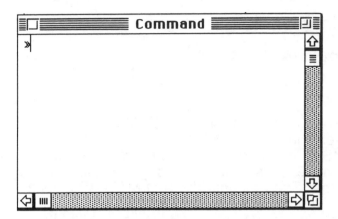

Figure 2.1 MATLAB Command Window.

```
»2*3
ans = 6
»3/5
ans = 0.6000
»3^8
ans = 6561
```

Parentheses work too, as the following dialog shows

```
»(3+5)*8
ans = 64
»3+5*8
ans = 43
```

From these dialog fragments you can see that MATLAB seems to turn your fairly expensive personal computer into an inexpensive calculator.

2.1.3 General Commands

MATLAB responds to each of your commands. When you type a characters or character combination that it understands, it performs the task appropriate to that command. Otherwise, MATLAB provides a diagnostic statement to let you know why your command is inappropriate. Table 2.1 lists a set of some general words that cause MATLAB to provide information and do some useful tasks.

Typing `help` makes MATLAB list all of the words and operators that you can use. Adding an operation or the name of a function after `help` gives information about that operation or function. For example

```
»help sin
SIN        SIN(X) is the sine of the elements of X
```

shows how helpful MATLAB can be. Of course, variable X is in radians. Try typing `Help` to see that MATLAB is case sensitive. The help documentation here is misleading, because `sin` not `SIN` is MATLAB's sine function.

The examples in Sec. 2.1.2 suggest existence of a variable having the name `ans`. When you type an expression without making an assignment, MATLAB assigns the result to the variable `ans`. Create a variable by typing a variable name before an assignment (=) symbol, followed by an expression. Variable

Word	Operation
help	Describes MATLAB's operators and functions
who	Lists names of variables
whos	Lists names of variables and tells their size
what	Lists M-Files on your disk
size	Returns the dimensions of its argument
length	Returns the maximum dimension of its argument
clear	Clears the workspace of all of your variables
quit	Terminates your MATLAB session
exit	Terminates your MATLAB session
save	Save your MATLAB workspace into a MATLAB MAT-File
load	Load a MATLAB MAT-File

Table 2.1 General Operations.

names start with any alphabetic character. After the first character in a name, you may use as many alphanumeric characters or the underscore character (_) as you wish, but MATLAB remembers only the first 19 characters.

Use who or whos to list the names of all variables in the workspace memory. whos gives more details than who, which just lists the names. size returns an array whose elements are the number of rows and columns in its matrix argument. length returns a scalar value equal to the length of its vector argument. We will say more about these commands in Sec. 2.3, which describes matrices and vectors. Typing clear removes all of your variables from the current workspace. Did you type save recently? Using save stores all current variables into a file having the name "MATLAB.MAT." Typing load recalls the contents of the most recent save to MATLAB.MAT.

2.1.4 Special Characters and Variables

MATLAB has special characters that you can use to control MATLAB or to form expressions. Table 2.2 gives a summary of these characters. The percent (%) symbol provides a way to document your work. If you type

```
»% This is a comment
```

Character	Function
[ ]	Delineate vectors and matrices
( )	Control operator precedence
,	Separate function or subscript arguments or separate statements on a command line
;	End rows in a matrix or suppress printing
%	Denote a comment
:	Vector or array generation
!	Execute a system function (MS-DOS) only

Table 2.2 MATLAB Characters.

Variable	Value
ans	Default assignment variable
eps	Value of floating-point precision
pi	Value of π
i or j	$\sqrt{-1}$
Inf	Infinity (∞)
NaN	IEEE definition of Not-a-Number
flops	Floating-point operation count
nargin	Number of function input arguments
nargout	Number of function output arguments

Table 2.3 MATLAB Variables.

MATLAB ignores the line and displays another prompt (»). The percent symbol can appear in-line following a statement. For example

```
»123 % This is a comment
ans = 123
```

Omission of the percent symbol produces an error message. Two or more statements can appear on a single line, if you separate each statement from its neighbors with a comma or a semicolon, as in

```
»123; 456, 789;
ans =    456
```

This example shows that the semicolon, in addition to serving as a statement separator, suppresses printing of the preceding statement. The left- and right-bracket characters ([ ]), comma (,), and colon (:) apply to vector or matrix generation, and Section 3.1 describes these. The exclamation symbol (!) tells MATLAB that the characters that follow are an operating system command (MS-DOS and UNIX only). For example, the line

```
»!edit myFile
```

invokes the MS-DOS editor, allowing you to edit or view the file "myFile." When you leave the editor control returns to MATLAB's Command window.

MATLAB defines special variables that appear in Table 2.3. You already know that ans stores the result of an expression that has no assignment. The default variable ans can appear in any expression, just as an ordinary variable that you define. Constant eps contains the value of MATLAB's floating-point precision. Variable pi stores the numerical value of π. Letters i or j represent $\sqrt{-1}$. Inf represents infinity, and NaN stands for *Not-a-Number*, an IEEE standard that results from operations without definition, such as 0.0/0.0. Variable flops accumulates the number of floating-point operations. flops(0) resets the count. Variables nargin and nargout contain the current number of input and output arguments of a function (See Sec. 2.4). You can redefine any of these variables. Using clear removes all user variables from the workspace, and except for eps, resets the definitions of the special variables. If you work with complex numbers be careful not to use i or j as index counters.

2.1.5 MATLAB Files

MATLAB works with three file types. The first type, an M-file, stores scripts and functions. M-files have the extension .M. M-files can be either scripts or functions. See Section 2.4 for details about the difference between these two. The second file type, a MAT-file, stores data. MAT-files have the suffix .MAT. The third type of file is a diary file, which records all command-window activity. Diary files have no particular suffix and are ordinary text files. Professional MATLAB also has MEX-files. These contain compiled C or FORTRAN functions that are accessible from within MATLAB.

2.1.6 Saving and Recalling Sessions

When working with MATLAB on your computer you may wish to turn the computer off, coming back later to resume the current session. To accomplish this goal, without having to key in all the variables that you may have been using, before turning your computer off type

```
»save
```

This command creates or overwrites a MAT-file in the current subdirectory or your MATLAB folder with the name "matlab.mat." Later, when you need to recall the workspace that exists in "matlab.mat," type

```
»load
```

Variables in "matlab.mat" overwrite variables of the same name in the current workspace, but variables with different names in the current workspace persist. To store and recall different sessions, use different names. For example

```
»save myFile
```

stores your current session in "myFile.MAT." To recall "myFile" type

```
»load myFile
```

To save only certain variables append those variables to the command. For example, to save var1 and var2 in "myFile" type

```
»save myFile var1 var2
```

Then

```
»load myFile
```

recalls just var1 and var2 into the current workspace.

2.1.7 Recording a MATLAB Session

To record your Command window session, use the diary command. This command has the syntax

```
diary [<File_Name>]
```

If you omit <File_Name> and MATLAB has no current diary file, then MATLAB opens the diary file with the name diary, creating it if necessary. Typing diary on turns diary recording on, and diary off turns recording off. If a current diary file already exists, typing only diary toggles diary recording on and off. Later, you can read the file with your word processor.

2.2 Scalars

The basic entity in MATLAB is a matrix having a distinct number of rows and columns. When either the number of rows or the number of columns reduces to one the matrix is an array or vector. With both the number of rows and columns equal to one the matrix becomes a scalar. This section explains how to create a scalar, and describes the operators and functions to use in their manipulation.

2.2.1 Creation of a Scalar Variable

Creation of a scalar variable simply involves writing an assignment statement. For example, typing

```
»k=1.38e-23
```

defines variable k, which you can use later in computations that involve Boltzmann's constant. In exponential notation you can use either lower-case e or upper-case E. MATLAB does not distinguish between floating-point and integer numbers. For example, when you type

```
»b=6.00
```

MATLAB responds with

```
b = 6
```

The `format` command gives some control of the appearance of MATLAB responses. The `format` commands appear in Table 2.4. Using some of these gives, for example

```
»format short
»6.2
ans = 6.2000
»format short e
»6.2
ans = 6.2000e+00
»format long
»6.2
ans = 6.20000000000000
```

Command	Effect
format short	Fixed-point, five digit
format long	Fixed-point, 15 digit
format short e	Floating-point, 5 digit
format long e	Floating-point, 15 digit
format hex	Hexadecimal
format +	+, -, and blank for positive, negative, and zero
format bank	Fixed-point, two digit
format compact	Suppresses an extra line feed
format loose	Includes an extra line feed

Table 2.4 Format Commands.

2.2.2 Arithmetic Operations

In addition to the basic arithmetic operations +, -, *, /, and parentheses in the examples of Section 1.2, MATLAB defines the left division operator (\ ). For example

```
»4\1
ans = 0.2500
```

shows that the left division operator divides the left operand into the right.

2.2.3 Relational and Logical Operators

Table 2.5 lists MATLAB's relational operators. For these the value that returns is either true (1) or false (0). For example

```
»4<5
ans = 1
»5<4
ans = 0
4<=4
ans = 1
»~ans
ans = 0
»0|1
ans = 1
»0&1
ans = 0
```

Any nonzero positive or negative number is logically true. Arithmetic operators take precedence over the relational operators. For example

```
»5+3>2
ans = 1
»5+(3>2)
ans = 6
```

Relational operators take precedence over logical operators, for example

```
»2&0.5>-0.4
ans = 1
```

Operator	Effect
<	Less than
<=	Less than or equal
>	Greater than
>=	Greater than or equal
==	Equal
~=	Not equal
&	Logical AND
|	Logical OR
~	Not

Table 2.5 Relational and Logical Operators.

Function	Comment
abs	Absolute value or magnitude of complex number
angle	Phase of complex number in radians
sqrt	Square root
real	Real part of complex number
imag	Imaginary part of complex number
conj	Complex conjugate of complex number
round	Round to nearest integer
fix	Round towards zero
floor	Round towards $-\infty$
ceil	Round towards ∞
sign	Signum (returns 1, 0, or -1 as argument is positive, zero, or negative)
rem	Remainder
exp	Base e = 2.7183 exponential
log	Natural logarithm
log10	Base 10 logarithm

Table 2.6 Basic Math Functions.

is the same as

```
»2&(0.5>(-0.4))
ans = 1
```

But

```
»0.5>(-.4)&2
ans = 1
»0.5>((-.4)&2)
ans = 0
```

2.2.4 Basic Functions

Table 2.6 lists basic math functions available. For example

```
rem(12,7)
ans = 5
»exp(1)
ans = 2.7183»fix(3.5)
ans = 3
»floor(3.5)
ans = 3
»ceil(3.5)
ans = 4
»round(3.5)
ans = 4
```

2.2.5 Trigonometry and Hyperbolic Functions

MATLAB has the usual trig and hyperbolic functions. Table 2.7 lists these. The arguments of trig functions and the return values of their inverses are in radians.

Function	Comment
sin	Sine
cos	Cosine
tan	Tangent
asin	Inverse sine
acos	Inverse cosine
atan	Inverse tangent
atan2	Four-quadrant inverse tangent (y,x)
sinh	Hyperbolic sine
cosh	Hyperbolic
tanh	Hyperbolic
asinh	Inverse hyperbolic sine
acosh	Inverse hyperbolic cosine
atanh	Inverse hyperbolic tangent

Table 2.7 Trig Functions.

Notice that the first argument of function `atan2` is the y component, the second is the x component.

2.2.6 Complex Numbers

MATLAB does complex numbers! In fact, all of the arithmetic, relational, and logical operators, basic, trig, and hyperbolic functions apply to complex number calculation. Logical operators regard a complex number with either real or imaginary part nonzero as logically true. Except for equality == and inequality ~=, the relational operators use the real part of their complex operands. The equality and inequality operators demand that the complex numbers be the same or different. MATLAB defines variable j or i as $\sqrt{-1}$. Be careful not to redefine either if you want to continue to use them to represent $\sqrt{-1}$. Using the command `clear` erases the workspace and restores the identity of j and i. With MATLAB 3.5 use j or i algebraically, not as tags. For example

```
»1+3*j
ans = 1.0000 + 3.0000i
»2+5i
??? 2+5i
        |
Missing operator, comma, or semicolon.
```

and

```
»y=sqrt((4+j*5)*(2+6*j)/(3-i*6))
y = 0.1359 - 2.4533i
»abs(y) % Magnitude
ans = 2.4570
»180*angle(y)/pi % Phase in degrees
ans = -86.8299
```

However, with MATLAB 4.X you can tag the imaginary part with i or j following immediately after the imaginary value. Repeating the previous example with MATLAB 4.X gives

```
»1+3j
ans = 1.0000 + 3.0000i
»2+5i
ans = 2.0000 + 5.0000i
»y=sqrt((4+5j)*(2+6j)/(3-6i))
y = 0.1359 - 2.4533i
»abs(y) % Magnitude
ans = 2.4570
»180*angle(y)/pi % Phase in degrees
ans = -86.8299
```

Do not attach the j or i tag prior to the imaginary value. If you wish, you can avoid the implication of multiplication and write complex numbers as in MATLAB 3.5.

2.3 Arrays and Matrices

MATLAB's utility for circuit analysis is due to its matrix processing capabilities. Among the various mathematics programs having matrix capability, MATLAB is probably the easiest to use, is highly accurate, and is stable in operation. For these reasons, MATLAB is effective for solving any circuit problem that you can formulate as a matrix computation.

2.3.1 Creation and Manipulation of Vectors and Matrices

Creation of vectors and matrices is the starting point of most MATLAB analyses. You create these manually with MATLAB statements or with functions, using M-files, or load them from MAT-files stored on a disk.

Manually

A simple way to create a vector or an array is to write a sequence of numbers, surrounding the sequence with square braces []. For example

```
»vec=[1 3 2 6 9]
vec = 1      3      2      6      9
```

To create an array of equally-spaced values use the colon operator as in a:b:c, where a is the starting value, b is the increment, and c is the final value. For example

```
»vec=0:2:10
vec = 0      2      4      6      8      10
```

or

```
»vec=-pi/2:pi/4:pi/2
vec = -1.5708   -0.7854   0   0.7854   1.5708
```

The function linspace(x1,x2,n) also creates a sequence of linearly-spaced points, but here the third argument n gives the number of points, x1 is the initial value, and x2 is the final value.

```
»vec=linspace(0,10,6)
»vec = 0      2      4      6      8      10
```

The final value can be smaller than the initial value. To create a sequence of equally-spaced logarithmic values use logspace(dec1, dec2, n). Here, `dec1` is the initial power of ten, `dec2` is the final power of ten, and `n` is the number of points. For example

```
»vec=logspace(1,2,5)
vec = 10.0000   17.7828   31.6228   56.2341   100.0000
```

A simple way to create a matrix in MATLAB is to enter the numerical values row by row, separating each number from its neighbor with one or more blanks or a comma and each row with a semicolon. Surround the entire set of numbers with square braces. For example

```
»A_Mat=[1 2 3;4 5 6;7 8 9]
A_Mat =
1       2       3
4       5       6
7       8       9
```

Starting each row on a new line, the carriage return replaces the semicolons, so

```
»A_Mat=[1 2 3
        4 5 6
        7 8 9]
A_Mat =
1       2       3
4       5       6
7       8       9
```

creates the same matrix and provides visual documentation of the completion of each row. If a matrix is too large to enter on a single line you can use a sequence of three periods to indicate continuation to the next line.

Functions

The matrix functions `ones`, `zeros`, or `eye` generate matrices containing all ones, all zeros, or ones on the diagonal with zeros off of the diagonal (i.e. an identity matrix). These functions can have either one or two integer arguments or a matrix argument. The one-integer form creates a square matrix having order equal to the integer. For the two-integer version, the number of rows equals the first integer and the number of columns equals the second. For example

```
»a=3;
»b=4;
»Iab=eye(a,b)
Iab =
1       0       0       0
0       1       0       0
0       0       1       0

»Ones_ab=ones(a,b)
Ones_ab =
1       1       1       1
1       1       1       1
1       1       1       1
```

```
»Zeros_ba=zeros(b,a)
Zeros_ba =
0       0       0
0       0       0
0       0       0
0       0       0
```

When the argument of ones, zeros, or eye is a matrix, the resulting matrix has the same size as the matrix argument, except when the matrix is a scalar. MATLAB 4.X considers the matrix argument form of these functions obsolete, and future versions will no longer support this usage. You can create diagonal matrices using diag(vec,k). Here, vec is an array of numbers and k is an integer giving the diagonal position. Omitting k or setting k equal to zero specifies the main diagonal. Setting k greater than zero moves the diagonal up, and making k less than zero moves the diagonal down. For example

```
»v=1:3;
»diag(v)
ans =
     1       0       0
     0       2       0
     0       0       3

»diag(v,1)
ans =
     0       1       0       0
     0       0       2       0
     0       0       0       3
     0       0       0       0
»diag(v,-1)
ans =
     0       0       0       0
     1       0       0       0
     0       2       0       0
     0       0       3       0
```

The function tril(mat,k) constructs a new matrix from mat using mat's values from the k-th diagonal downwards. For example

```
»tril(Ones_ab)
ans =
     1       0       0       0
     1       1       0       0
     1       1       1       0
»tril(Ones_ab,1)
ans =
     1       1       0       0
     1       1       1       0
     1       1       1       1
»tril(Ones_ab,-1)
ans =
     0       0       0       0
     1       0       0       0
     1       1       0       0
```

Function `triu(mat, k)` creates an upper-triangular matrix, commencing at the k-th diagonal.

M-Files

M-files are text files that contain MATLAB statements. When you type the name of an M-file, MATLAB reads and sequentially executes the statements. You can write M-files that have matrix generation statements that define any matrix that you want to create. For example, running the M-file Ex_2_2.M shown in Fig. 2.2 from the Command window gives

```
»EX_2_2
A_Mat =
        1       2       3
        4       5       6
        7       8       9
```

The advantage of this process over manual entry in MATLAB's Command window is that you can edit the M-file to correct errors.

The M-file has to be in the path or working folder so that MATLAB can find the file. Use the `matlabpath` command and your PC's path naming convention to tell MATLAB where to look for your M-files. MATLAB 4.X still responds to `matlabpath`, but the documentation shows that this command is now `path`. On a Macintosh, **Set Path...** is an option on the **M-File** menu in MATLAB 3.5 and on the **File** menu in MATLAB 4.X. M-files also provide complete scripts for automating entire MATLAB sessions and for function creation. Section 2.4.1 describes this use of M-files.

```
% Matrix Creation
A_Mat=[1, 2, 3
       4, 5, 6
       7, 8, 9]
```

Figure 2.2 M-File Ex_2_2.M.

MAT-Files

Section 2.1.6 describes how to save a MATLAB session and later recall the session for further work using the `save` and `load` commands. Using a word processor, you also can create a MAT-file that has data for a matrix and load this data into your current session with the `load` command. Each line of the MAT-file provides a row of data for the matrix, with spaces separating numbers. Each line terminates with a carriage return. To load the data, from the Command window type

```
load <File_Name>
```

On a Macintosh <File_Name> can have the suffix .MAT as in "myFile.mat," but the suffix is unnecessary. If the suffix is present you do not have to give the suffix as part of <File_Name> in the load command. With an MS-DOS computer you can load a text file not having a .MAT extension by entering the file name and appending a period. For example, "myFile." loads "myFile" although this file has no .MAT extension. When the load command executes, MATLAB creates a matrix having the same name as the file name. Although upper- and lower-case characters are not significant in the file name, they are in MATLAB. So typing

```
load myFile
```

and

```
load MyFile
```

create different matrices that have the same numerical values.

2.3.2 Subscript Reference

Reference to different elements of a vector or matrix in MATLAB uses parentheses, integers, and commas as in most programming languages (but not the C language). For example, the third row, second column element of matrix my_Mat is my_Mat(3,2). The row and column integers can be expressions and values that round to the nearest integer. in addition, arrays can replace the row or column indices to select specific sets of rows or sets of columns from a matrix. For example

```
»a=diag(1:5)
a =
     1      0      0      0      0
     0      2      0      0      0
     0      0      3      0      0
     0      0      0      4      0
     0      0      0      0      5


»b=a([2,4],[2 4 5])
b =
     2      0      0
     0      4      0
```

The array [2 4] selects the second and fourth rows, the array [2 4 5] selects the second, fourth, and fifth columns, so matrix b is a 2×3 matrix that consists of the second and fourth rows and the second, fourth, and fifth columns of a. The colon operator is useful to create row- or column-array indices. By itself, the colon operator indicates all of the row or columns. For example, with matrix a above

```
»b=a(:,[2,4])
b =
     0      0
     2      0
     0      0
     0      4
     0      0
```

Array indexing applies to both the left- and right-hand side of an assignment.

For example

```
»a(:,[2 4])=a(:,[4,2])
a =
      1      0      0      0      0
      0      0      0      2      0
      0      0      3      0      0
      0      4      0      0      0
      0      0      0      0      5
```

interchanges the second and fourth columns of matrix a.

2.3.3 Arithmetic Operations

The addition (+) and subtraction (-) operators add and subtract their operand matrices. Multiplication (*), left (\) and right division (/), and the exponential operator (^) have array and matrix interpretations. Preceding these operators with a period gives these operators array interpretations. For example

```
»a=1:5;
»b=2:6;
»a.*b
c =
      2      6     12     20     30
```

The operand arrays or matrices need to have the same dimensions. Consider the MATLAB 3.5 dialog

```
»3.\c
ans =
     0.6667     2.0000     4.0000     6.6667    10.0000
»3./c
??? Error using ==> /
Matrix dimensions must agree.

»3.0./c
ans =
     1.5000     0.5000     0.2500     0.1500     0.1000
»(3)./c
ans =
     1.5000     0.5000     0.2500     0.1500     0.1000
```

The MATLAB 3.5 response to the 3.\c statement divides each element of array c by the number 3.0. The problem with the 3./c statement is that MATLAB 3.5 does not know whether we intend array right division or matrix right division. Either 3.0./c or (3)./c tells MATLAB to do array right division. For 3.\c, 3.*c, 3.+c, or 3.-c MATLAB knows that you intend the scalar operation or that there is no difference. But 3.^c is not the same as 3.0.^c or (3).^c. MATLAB 4.X associates the decimal with the / operator.

2.3.4 Relational and Logical Operators

The relational and logical operators of Table 2.5 apply where the operands are arrays or matrices having the same dimensions. The operators affect the individual elements on a term-by-term basis and produce ones or zeros according to the validity or invalidity of the comparison or logical operation. Functions any and all return ones or zeros according to whether any or all of the arguments of an array are nonzero. If these functions take a matrix as an argument they return an array of ones or zeros according to whether the columns of the matrix have any or all nonzero entries.

2.3.5 Elementary and Trig Functions

The elementary and trig functions of Tables 2.6 and 2.7 can have array or matrix arguments. These functions return an array or matrix with elements that are the return value of the function for each element of the argument. The example

```
»x=(-pi:pi/2:pi)';
»[x sin(x)]
ans =
    -3.1416    -0.0000
    -1.5708    -1.0000
          0          0
     1.5708     1.0000
     3.1416     0.0000
```

creates a 5×2 matrix that pairs angles in radians from -π to π with the corresponding sine-function values. In the line that defines x the apostrophe (') converts the row vector to a column vector.

2.3.6 Complex Numbers

Any array or matrix element can be a complex number. Try entering

```
»c_mat=[1 +j*2 1 -j*2; 2 +j*2 2 -j*1]
c_mat =
    1.0000    0 + 2.0000i    1.0000    0 - 2.0000i
    2.0000    0 + 2.0000i    2.0000    0 - 1.0000i
```

This result may surprise you. The problem here is that extra blanks change an apparently single complex number into two complex numbers. For example, 1 +j*2 becomes 1+j0 and 0+j2. To create the 2×2 matrix that we intend, type

```
»c_mat=[1+j*2 1-j*2; 2+j*2 2-j*1]
c_mat =
    1.0000 + 2.0000i    1.0000 - 2.0000i
    2.0000 + 2.0000i    2.0000 - 1.0000i
```

Another way to enter this matrix is to form the real and imaginary parts as separate matrices and add these to compute the result.

```
»c_mat=[1 1;2 2]+j*[2 -2; 2 -1]
c_mat =
    1.0000 + 2.0000i    1.0000 - 2.0000i
    2.0000 + 2.0000i    2.0000 - 1.0000i
```

Two matrix transpose operators exist. The apostrophe (') is a post-fix operator that creates the complex-conjugate transpose, so

```
»c_mat'
ans =
    1.0000 - 2.0000i    2.0000 - 2.0000i
    1.0000 + 2.0000i    2.0000 + 1.0000i
```

To obtain the transpose without taking the complex conjugate of each element, use the period-apostrophe post-fix operator (`.'`). Using this combination gives

```
»c_mat.'
ans =
    1.0000 + 2.0000i    2.0000 + 2.0000i
    1.0000 - 2.0000i    2.0000 - 1.0000i
```

Be careful to use the appropriate transpose operator.

Matrix Addition (+) and subtraction (−) have the same result as for their array interpretations. The matrix multiplication operator (*) multiplies the second operand by the first. The i-th row, j-th column element of the product is the sum of successive products using the i-th row elements in the first matrix and the j-th column elements of the second. For example

```
»a_mat=[1 2;  3  4];
»b_mat=[4 3;  2 1];
»disp(a_mat*b_mat)
     8       5
    20      13
```

The second-row, first-column element results from the computation

$$3 \times 4 + 4 \times 2 = 12 + 8 = 20$$

Matrix division does not exist. However, MATLAB defines left- and right-division operators for matrices Their interpretation is given in the next section. In the above example function `disp` displays the value of its argument. The argument can be text, but the text form usually appears in an M-file (See Section 2.4.1) to document work in the Command window.

2.3.7 LU Factors and Inversion

The matrix function `inv` takes the inverse of its matrix argument. Let's explore the `inv` function with MATLAB V3.5

```
»null_mat = []
null_mat = []
»inv(null_mat)


»scalar_mat = 3
scalar_mat = 3
»inv(scalar_mat)
ans = 0.3333
```

A null matrix is not much, but it exists. In MATLAB 3.5 the inverse of a null matrix does not appear to exist, although MATLAB gives no warning of a problem. However, MATLAB 4.X gives `ans = [ ]`. A single-element matrix is the same as a scalar. Of course, the inverse of a scalar is the reciprocal of its value. MATLAB gives the inverse of matrix `b_mat` in the example above as

```
»inv(b_mat)
ans =
   -0.5000     1.5000
    1.0000    -2.0000
```

You can verify this inverse of matrix b_mat using the adjoint and determinant from Section 1.2. MATLAB calculates the inverse using the L and U factors (See Section 1.2.2). If

```
»x=[0;1];
```

calculating y from

```
»y=inv(b_mat)*x
y =
     1.5000
    -2.0000
```

is not as efficient as computing y from the L and U factors directly. Use the left-division operator (\) to obtain $B^{-1}X$. For example

```
»y=b_mat\x
y =
     1.5000
    -2.0000
```

The result is the same as before, but the second method is simpler to type and gives MATLAB an easier task. Left-to-right operator precedence applies here, so

```
»y=a_mat*b_mat\x
y =
    -1.2500
     2.0000
```

and

```
»y=(a_mat*b_mat)\x
y =
    -1.2500
     2.0000
```

are the same, but

```
»y=a_mat*(b_mat\x)
y =
    -2.5000
    -3.5000
```

is different. Essentially, y=b_mat\x solves the matrix equation

$$b\_mat * y = x$$

while y=x/b_mat solves the equation

$$y * b\_mat = x$$

2.4 M-File Scripts and Functions

Instead of writing single-line commands in MATLAB's Command window, you can write text files that contain statements that MATLAB understands. You can write these text files within the MATLAB environment, using MATLAB's own simple text window or you can prepare these using your own word processor. If you use your own word processor, save the file as an ASCII text file with carriage returns at the end of each line, giving the file name a .M extension. There are two kinds of M-files. The first kind is a script that MATLAB runs, as if the commands were to occur one-at-a-time in the Command window. The second kind of M-file is a function file. A function file defines a new MATLAB function.

2.4.1 Script M-Files

Use scripts when you want to automate your work. Write a sequence of MATLAB statements to accomplish your task. Do this at your leisure, getting every statement correct and in the right order. When your script is done, run it from the Command window and inspect the results. Some useful MATLAB statements to use in scripts, although they also apply to the Command window, are the `if`, the `for`, and the `while` statement.

The syntax of an `if` statement is

```
if expression_1
      statements
elseif expression_2
      statements
else
      statements
end
```

The first group of statements execute if `expression_1` is true (nonzero). `expression_1` can be an array or matrix, in which case it is true if and only if all of its elements are true. The second group of statements execute if `expression_1` is false and `expression_2` is true. The final group execute when both `expression_1` and `expression_2` are false. The `elseif` construct can occur more than once. You can omit the `elseif` or the `else` block if you wish.

The form of the `for` statement is

```
for variable = expression
      statements
end
```

In general, `expression` is a matrix. `Variable` assumes successive columns of this matrix and the statements execute for each column of the expression.

The `while` statement is

```
while expression
      statements
end
```

Here, the statements execute while all elements of expression are true.

```
for ii=1:5
      for jj=1:3
            if ii>jj
                  amat(ii,jj)=ii*jj;
            elseif ii<jj
                  amat(ii,jj)=ii/jj;
            else
                  amat(ii,jj)=0;
            end
      end
end
disp('Matrix amat is ...')
disp(amat)
```

Figure 2.3 M-File EX_2_3.M.

Figure 2.3 shows an example of an M-file script. Running this script gives

```
»EX_2_3
Matrix amat is ...
         0      0.5000      0.3333
    2.0000           0      0.6667
    3.0000      6.0000           0
    4.0000      8.0000     12.0000
    5.0000     10.0000     15.0000
```

2.4.2 Function   M-Files

MATLAB interprets M-files having a first line that begins with the word
function as a function file. The syntax of a function file is

```
function <Return_Var> = ...
      <Function_Name>(<Arg1>, [<Arg2>, ...])
statements
```

Here, <Return_Var> names a variable known only within the function and
stores the result of the function. <Function_Name> names the function. Usually
you name the M-file with the name <Function_Name>.M. Although choosing
the M-file name to be the same as the function name is unnecessary, doing
otherwise may be confusing. Within parentheses, list the function's arguments.
The arguments are local to the function and cannot conflict with any variables
that the Command window knows about. For example, you can use the names i
or j as integer arguments in a function, without affecting use of i or j as $\sqrt{-1}$ in the
Command window. At least one of the statement(s) has to assign an expression
to <Return_Var>. The last assignment to <Return_Var> defines what the
function returns. A function can return more than one variable. In this case, the
syntax of the function is

```
function [<Return_Var_1>, <Return_Var_2>,...] = ...
      <Function_Name>(<Arg1>,[<Arg2>, ...])
statements
```

Type the first set of square braces. The second set of square braces indicate
that <Arg2> and any subsequent arguments are optional.

```
function nrows=rows(aMat)
% Returns the number of rows in aMat.
[nrows,ncols]=size(aMat);
```

Figure 2.4 New MATLAB Function `rows`.


```
function ncols=cols(aMat)
% Returns the number of columns in aMat.
[nrows,ncols]=size(aMat);
```

Figure 2.5 New MATLAB Function `cols`.

Figures 2.4 and 2.5 show examples of MATLAB functions. The functions `rows` and `cols` compute the number of rows and the number of columns of a matrix. Since `size` computes both of these values, why write functions to retrieve each value separately? Because, the new functions allow you to write a single-line statement using either the number of rows or the number of columns. For example

```
»amat=zeros(3,5);
»ones(rows(amat),cols(amat))
ans =
        1       1       1       1       1
        1       1       1       1       1
        1       1       1       1       1
```

To obtain this result using `size` requires an extra line

```
»amat=zeros(3,5);
»[nrows,ncols]=size(amat)
»ones(nrows,ncols)
ans =
        1       1       1       1       1
        1       1       1       1       1
        1       1       1       1       1
```


2.5 Graphing with MATLAB

To see a plot of your calculations as an independent variable changes and with a set of parameter values gives a feeling to your results that you cannot gain just by looking at the equations. The operative word for graphs is `plot`. To see how `plot` works in MATLAB 3.5, try

```
»x=-pi:pi/50:pi;
»y=sin(x);
»axis([-pi pi -1 1])
»plot(x,y)
```

Figure 2.6 shows the plot that appears in MATLAB's graph window. The first argument of `plot`, in this case `x`, is the independent variable. The second argument `y` is the dependent variable. If you use far fewer points than 50 in the row arrays, then the curve begins to show the breaks as the graph actually is a sequence of straight lines that connect successive pairs of array values. If `plot`

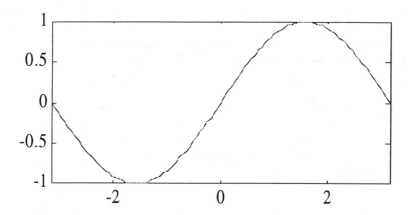

Figure 2.6 MATLAB Plot of the Sin Function.

has only one array argument, then the elements of that array plot as functions of the index values. If either variable x or y becomes a matrix and the other remains a row or column array, then the matrix rows or columns plot as functions of the array elements. Notice that `axis` appears before the `plot` command. The new object-oriented handling of graphics in MATLAB 4.X requires that you use `axis` after the `plot` statement. See your program documentation for details about `plot` and `axis`.

Without the `axis` statement, the axis scales automatically. The components of the four-component vector argument of `axis` are the minimum and maximum of the x scale and the minimum and maximum of the y scale. Using `axis` with no argument holds the current graph's axis scales for subsequent plots. Typing `axis` again returns to the auto scaling mode. With `axis('square')`, the y-axis to x-axis aspect ratio is 1:1. Circles appear as circles. Using `axis('normal')` returns the aspect ratio to normal.

Other plot commands include

```
plot(y)
plot(x,y,'<line_type>')
plot(x1,y1,x2,y2,...)
semilogx(x,y)
semilogy(x,y)
loglog(x,y)
polar(θ,ρ)
```

Line		Point		Color	
Solid	-	Point	.	Red	r
Dash	--	Plus	+	Green	g
Dot	:	Star	*	Blue	b
Dash-dot	-.	Circle	o	White	w
		X-mark	x	Invisible	i

Table 2.8 Line-Type Characters

The simple form `plot(y)` plots the components of array y as a function of the index. You have seen that `plot(x,y)` plots array y vs. array x when both y and x are arrays. When x or y is a matrix and the other is an array, if the number of rows of the matrix equals the number of array elements, then the columns of the matrix plot versus the array. If the number of columns of the matrix equals the number of array elements, then rows of the matrix plot versus the array elements. With equal numbers of rows and columns, the matrix rows plot versus the array elements. MATLAB flags any other situation as an error. If x and y are matrices of the same size, then columns of x plot versus columns of y. Using <line_type> specifies a line, point, or color. Table 2.8 shows the symbols to use in place of <line_type> for various line, point, or color choices. If you want each line of a multi-line plot to have a different line, point, or color, use the `plot(x1,y1,'<line_type_1>',x2,y2,'<line_type_1>',...)` form of the `plot` command.

With `semilogx(x,y)`, `semilogy(x,y)`, and `loglog(x,y)` you can create y vs. log(x), log(y) vs. x, and log(y) vs. !og(x) logarithm (base 10) plots. The `polar(q,r)` command places a polar plot in the graph window. The angle θ is in radians and ρ is the magnitude.

To enhance your plot with axis labels, a title, annotation text, or a grid use the following commands

```
title('<Title>')
xlabel('<x_Text>')
ylabel('<y_Text>')
text(x,y,'<Text>',['sc'])
gtext('<Text'>)
grid
```

With MATLAB 3.5, when you include `sc` in the text command, the x and y coordinates are in screen coordinates, where (0,0) is the lower left-hand corner of the screen and (1,1) is the upper right-hand corner. For example, to improve the display of the sine function in Fig. 2.6, run the following sequence of commands to obtain the new plot in Fig. 2.7.

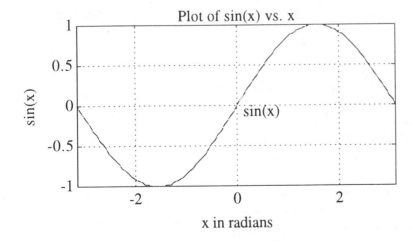

Figure 2.7 Graph with Annotation.

```
»x=-pi:pi/50:pi;
»y=sin(x);
»axis([-pi pi -1 1])
»plot(x,y)
»xlabel('x in radians')
»ylabel('sin(x)')
»title('Plot of sin(x) vs. x')
»text(0.55,0.5,'sin(x)','sc')
»grid
```

In MATLAB 4.X, the `sc` coordinate option is unavailable. Instead, use the `axes` command to create a coordinate system where the horizontal and vertical axes align with the bottom and left side of the window. MATLAB 4.X users refer to a window as a figure. Refer to the MATLAB 4.X documentation for details.

Placing text with the `text` command takes some practice. The gtext command permits you to place text visually on the graph by showing a crosshair on the screen. Position the crosshair using the mouse. When the crosshair is where you want your text to appear, press the mouse button. Even using `gtext` takes some practice. When you want the text to appear to the left of an object, you have to anticipate the length of your text to the right of the crosshair.

You can measure points on a graph using various forms of the `ginput` command. For example, after entering

```
»ginput
```

in the Command window, click on the Graph window with the mouse button to make the Graph window active, position the cursor, and click the mouse button again. Position the cursor and click again as often as you like. Now type the carriage return key <CR>. The Command window becomes the active window and a list of the Y values that you clicked appears. To be more selective, use

```
»[xVal,yVal]=ginput(n);  <CR>
```

where n is an integer. The Graph window becomes active for n mouse clicks and then the Command window becomes active. `xVal` and `yVal` now are n-

Command	Action
shg	Show graph window (3.5)
clg	Clear graph window (3.5)
clc	Clear Command window
Any key	Bring Command window front
home	Home command cursor
figure	Create figure (4.X)
clf	Clear figure (4.X)
close	Close figure (4.X)
gcf	Get handle for current figure (4.X)
gca	Get handle for current axis (4.X)

Table 2.9 Command Window and Graph Window Commands.

component arrays having the coordinates of the n click points. Using `sc` as an argument of `ginput` with MATLAB 3.5 returns the click points in screen coordinates. These screen values are useful when using `gtext` to place text on a graph.

To superimpose additional graphs on a plot already in place use `hold [on]`. To allow plots to replace previous plots type `hold [off]`. Without `on` or `off` the `hold` command toggles the state. Table 2.9 shows some commands that allow you to affect the command and graph windows.

If you are using MATLAB 4.X with a color screen, you will appreciate the default black background of figures. Plots in this format automatically print with a white background. On a black and white screen you may want to have your figures appear with a white background. Use `whitebg` to change the background of your current figure and any subsequent figures that you create. If you want to start every session with figures having a white background, write an M-file that includes `whitebg` and any other commands you wish to issue when MATLAB starts. Name this file `startup.m`. MATLAB 4.X runs this file when it starts.

Chapter 3
Node and Mesh Analysis

This chapter describes three ways to formulate a set of linearly-independent equations for a linear circuit. The first method, nodal analysis, gives the circuit solution in terms of a set of linearly-independent node-voltage variables. The second method is mesh analysis, which finds the solution for a planar circuit using a set of linearly-independent mesh-current variables. The third method is modified nodal analysis (MNA). Many computer-aided circuit analysis programs use MNA. Modified nodal analysis uses node voltages as a portion of the set of solution variables. The rest of the solution set consists of currents for elements that do not have a natural nodal analysis representation or element currents that you want to appear in the set of solution variables. A voltage-controlled voltage source (VCVS), a current-controlled current source (CCCS), or a current-controlled voltage source (CCVS) are elements that have no natural nodal analysis representation. Other examples of this type of element are the ideal op amp, inductances with mutual coupling, or an ideal transformer.

This chapter has fifteen examples that show how to write circuit equations in matrix form and to solve these with MATLAB. The examples exist as M-files in the CK_EX subdirectory, or folder in Macintosh terms, and use function M-files that are in the CK_TOOL subdirectory. You can find these subdirectories on the disk that comes with this Manual. Appendix A lists the new M-file functions.

3.1 Nodal Analysis

Nodal analysis formulates a circuit's solution using a linearly-independent set of Kirchhoff current law equations, written in terms of the node voltages. Using the values of these node voltages and the values of the independent sources, you can determine the values of any remaining circuit variables. We assume that the circuit is proper. A proper circuit can have no voltage-source or inductance loops and no current-source or capacitance cutsets. The problem with a voltage-source or inductance loop is that the current in the loop has no unique value. Current-source or capacitance cutsets create separate circuits that have no unique voltage between a node in one part and any node in the other part.

3.1.1 Nodal analysis with Independent Current Sources

Nodal analysis of a linear circuit having n+1 nodes involves selecting a reference node and writing a set of n independent Kirchhoff Current Law (KCL) equations at each of the remaining nodes in terms of the node voltages. These node voltages constitute a linearly-independent set of solution variables. If the circuit has only conductances and current sources, you can write the independent set of equations in matrix form directly by inspection of the circuit. This fact arises directly from the summation form of KCL. Each conductance in the circuit connects between two nodes and carries a current proportional to both the node-voltage difference and the conductance value. If the sequence of the node voltages in the node-voltage solution vector is the same as for the KCL equations, then the node equations have diagonal symmetry. For example, if a conductance G_A connects from node i to node j then the terms shown in Eq. 3.1 appear in the KCL equations

$$\vdots$$

$$i \qquad \ldots + G_A\left(V_{N_i} - V_{N_j}\right) + \ldots = \ldots$$

$$\vdots$$ (3.1)

$$j \qquad \ldots + G_A\left(V_{N_j} - V_{N_i}\right) + \ldots = \ldots$$

$$\vdots$$

The coefficient G_A appears as a positive entry in the KCL equations in the i-th and j-th diagonal positions and as a negative entry in the i-th row and j-th column and in the j-th row and i-th column. If a conductance connects between a node i and the reference node, then that conductance appears only as a positive term in the i-th diagonal position.

Example 3.1 Independent Current Source Nodal analysis Example

For the circuit shown in Fig. 3.1

a) write the KCL node-voltage equations in terms of the individual conductance currents and node voltages V_{N1}, V_{N2}, and V_{N3}.

b) Rewrite the equations collecting coefficients common to each node voltage.

c) Express the three KCL node-voltage equations in matrix form. Notice the diagonal symmetry of the conductance matrix and the location of nonzero entries in the current source matrix.

d) Write the conductance matrix as the sum of five separate matrices, each involving only one conductance, and the current source matrix as the sum of two matrices, each involving only one of the current sources.

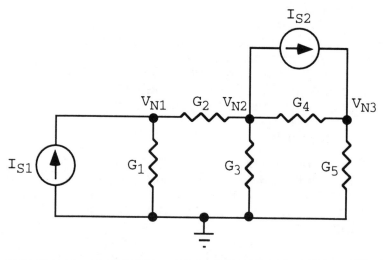

Figure 3.1 Circuit for Example 3.1.

Solution

a) The three KCL equations are

$$G_1 V_{N1} + G_2 \left(V_{N1} - V_{N2} \right) = I_{S1}$$

$$G_2 \left(V_{N2} - V_{N1} \right) + G_3 V_{N2} + G_4 \left(V_{N2} - V_{N3} \right) = -I_{S2}$$

$$G_4 \left(V_{N3} - V_{N2} \right) + G_5 V_{N3} = I_{S2}$$

b) Collecting coefficients of each node voltage

$$\left(G_1 + G_2 \right) V_{N1} - G_2 V_{N2} = I_{S1}$$

$$-G_2 V_{N1} + \left(G_2 + G_3 + G_4 \right) V_{N2} - G_4 V_{N3} = -I_{S2}$$

$$-G_4 V_{N2} + \left(G_4 + G_5 \right) V_{N3} = I_{S2}$$

c) Writing these in matrix form

$$\begin{bmatrix} G_1 + G_2 & -G_2 & 0 \\ -G_2 & G_2 + G_3 + G_4 & -G_4 \\ 0 & -G_4 & G_4 + G_5 \end{bmatrix} \begin{bmatrix} V_{N1} \\ V_{N2} \\ V_{N3} \end{bmatrix} = \begin{bmatrix} I_{S1} \\ -I_{S2} \\ I_{S2} \end{bmatrix}$$

You can write this matrix solution directly by inspection of the circuit. Each term on the diagonal consists of the sum of conductances that connect to each node. These terms are *self-conductance* terms. Each off-diagonal *mutual-conductance* term is the negative sum of conductances that connect between node pairs.

d) Rewriting the conductance and current source matrices as separate terms gives

$$\left\{ \begin{bmatrix} G_1 & 0 & 0 \\ 0 & 0 & 0 \\ 0 & 0 & 0 \end{bmatrix} + \begin{bmatrix} G_2 & -G_2 & 0 \\ -G_2 & G_2 & 0 \\ 0 & 0 & 0 \end{bmatrix} + \begin{bmatrix} 0 & 0 & 0 \\ 0 & G_3 & 0 \\ 0 & 0 & 0 \end{bmatrix} + \begin{bmatrix} 0 & 0 & 0 \\ 0 & G_4 & -G_4 \\ 0 & -G_4 & G_4 \end{bmatrix} \right.$$

$$\left. + \begin{bmatrix} 0 & 0 & 0 \\ 0 & 0 & 0 \\ 0 & 0 & G_5 \end{bmatrix} \right\} \begin{bmatrix} V_{N1} \\ V_{N2} \\ V_{N3} \end{bmatrix} = \left\{ \begin{bmatrix} I_{S1} \\ 0 \\ 0 \end{bmatrix} + \begin{bmatrix} 0 \\ -I_{S2} \\ I_{S2} \end{bmatrix} \right\}$$

This final form of the KCL matrix equation shows how each circuit element affects the structure of the node analysis matrix equation. A conductance that connects between a node and the reference node appears only in the diagonal position for that node. A conductance connecting between two nodes, neither of which is a reference node, appears in four positions. The two diagonal entries are positive, and the two off-diagonal terms are negative. The off-diagonal entries occur in the row of one node and the column of the other node.

For a current source, enter the current-source value into one or two rows of the excitation matrix on the right-hand side of the equation. The entry is positive in the row corresponding to the node that the current's direction arrow heads towards. In the row corresponding to the node that the direction arrow leaves, the entry is negative. Omit the entry if the node is the reference node.

The observations in Example 3.1 show that the conductance matrix for nodal analysis consists of an accumulation of terms for each conductance. Appendix A lists a new MATLAB function nmAcc to do this accumulation. Figure 3.2 shows the help documentation for nmAcc. Function nmAcc has four arguments. The first two, i and j, identify the conductance nodes. The third argument, val, is the conductance value. The last argument, bmat, identifies the conductance matrix onto which the conductance value accumulates. If either i or j is zero then accumulation of the conductance value occurs only onto the diagonal position of the nonzero node. When both i and j are nonzero, then positive accumulation of the conductance's value occurs onto each diagonal position and negative accumulation occurs onto each off-diagonal position. No accumulation occurs if both i and j are zero, which is an error.

```
amat=nmAcc(i,j,val,bmat)
 Accumulates G or R values "val" onto bmat.
 i and j are node (mesh) indices.
```

Figure 3.2 Function nmAcc Help.

```
amat=srcAcc(i,j,val,bmat)
    Accumulates Is or Vs values onto source matrix bmat.
      Node i is "from" node (drop mesh).
      Node j is "to" node (rise mesh).
```

Figure 3.3 Function srcAcc Help.

Current sources appear in the KCL equations as entries into the right-hand side excitation matrix. A current I_S having direction from node i to node j accumulates positively on the j-th node's row and negatively on the i-th node's row. If either node is the reference node then accumulation occurs only on the non-reference node. Appendix A lists a new MATLAB function srcAcc that accumulates a current source onto a current-source matrix. help documentation for srcAcc appears in Fig. 3.3. The first argument of srcAcc is the *from* node i, the node from which the source flows. The second argument is the *to* node j that the source flows towards. The third argument val gives the current-source value, and the fourth identifies the source matrix. Example 3.2 below illustrates use of these functions

Example 3.2 Resistances of a One-Ohm Cube

The circuit shown in Fig. 3.4 is a 1-Ω cube. Evaluate

a) The body-diagonal resistance between nodes 6 and 0.

b) The face-diagonal resistance between nodes 2 and 0.

c) The edge resistance between nodes 1 and 0.

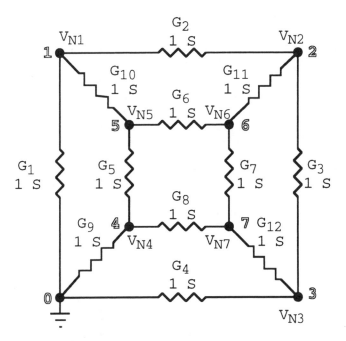

Figure 3.4 One-Ohm Cube.

Solution

Apply a 1-A source between each appropriate node pair and compute the voltage response between that node pair. The value of the voltage response is

```
% Example 3.2 - Resistances of a One-Ohm Cube
clear
% Define gMat
gMat=nmAcc(1,0,1,zeros(7));      gMat=nmAcc(2,1,1,gMat);
gMat=nmAcc(3,2,1,gMat);          gMat=nmAcc(0,3,1,gMat);
gMat=nmAcc(5,4,1,gMat);          gMat=nmAcc(6,5,1,gMat);
gMat=nmAcc(7,6,1,gMat);          gMat=nmAcc(4,7,1,gMat);
gMat=nmAcc(4,0,1,gMat);          gMat=nmAcc(5,1,1,gMat);
gMat=nmAcc(6,2,1,gMat);          gMat=nmAcc(7,3,1,gMat)
% A general current source for all three cases.
Is(:,1)=srcAcc(0,6,1,zeros(7,1));
Is(:,2)=srcAcc(0,2,1,zeros(7,1));
Is(:,3)=srcAcc(0,1,1,zeros(7,1));
% The voltage vectors for all three cases.
V=gMat\Is
disp('The body-diagonal resistance is:')
Rbd=V(6,1)
disp('The face-diagonal resistance is:')
Rfd=V(2,2)
disp('The edge resistance is:')
Re=V(1,3)
```

Figure 3.5 MATLAB M-file to Compute Body-Diagonal, Face-Diagonal, and Edge Resistances of 1-Ω Cube.

```
gMat =
     3    -1     0     0    -1     0     0
    -1     3    -1     0     0    -1     0
     0    -1     3     0     0     0    -1
     0     0     0     3    -1     0    -1
    -1     0     0    -1     3    -1     0
     0    -1     0     0    -1     3    -1
     0     0    -1    -1     0    -1     3
V =
    0.3333    0.3750    0.5833
    0.5000    0.7500    0.3750
    0.3333    0.3750    0.2083
    0.3333    0.2500    0.2083
    0.5000    0.3750    0.3750
    0.8333    0.5000    0.3333
    0.5000    0.3750    0.2500
The body-diagonal resistance is:
Rbd = 0.8333
The face-diagonal resistance is:
Rfd = 0.7500
The edge resistance is:
Re = .5833
```

Figure 3.6 MATLAB Solution of 1-Ω Cube Problem.

the same as the value of the resistance.

a) Apply a 1-A current source from node 0 to node 6 and measure the voltage at node 6.

b) Apply a 1-A current source from node 0 to node 2 and measure the voltage at node 2.

c) Apply a 1-A current source from node 0 to node 1 and measure the voltage at node 1.

The MATLAB M-file in Fig. 3.5 finds the solutions using a three-column excitation matrix. Function nmAcc accumulates the conductance values onto the conductance matrix g_{Mat}. Function srcAcc accumulates the source onto matrix I_s. Each column of I_s is a current source vector for either part a), b), or c). The node-voltage solution V is a 7×3 matrix, where each column of V is the node-voltage solution vector for the corresponding column of I_s. For example, voltage component V(6,1) is the voltage at node 6 when a 1-A current source connects from node 0 to node 6. This voltage numerically equals the body-diagonal resistance. The numerical results are in Fig. 3.6. The resistance values are 0.8333 Ω, 0.75 Ω, and 0.5833 Ω for the body-diagonal, face-diagonal, and edge resistances. These values agree with the 5/6 Ω, 3/4 Ω, and 7/12 Ω resistance values that you can determine from symmetry considerations.

3.1.2 Supernode Nodal Analysis

Existence of a voltage source in a circuit both simplifies and complicates the nodal analysis method. Each voltage source reduces the number of unknown node voltages by one. The number of independent KCL equations reduces to

$$\text{No. of Independent KCL Equations} = n - n_{vs} \qquad (3.2)$$

where n_{vs} is the number of voltage sources in the circuit. A voltage source that connects between a node and the reference node identifies that node's voltage. A voltage source connecting two non-reference nodes identifies one of the node voltages in terms of the other, eliminating one node voltage as an unknown in the solution. At the same time, each voltage source in a circuit has a current that you may wish or need to determine. Assume that finding any voltage source current is not of immediate concern. Then each voltage source eliminates one KCL equation from the set of KCL equations giving the node-voltage solution. Because each of the remaining equations is essential to complete the set of linearly-independent KCL equations necessary to determine the independent node voltages, we call these the *essential* KCL equations..

For a voltage source that connects between a node and the reference node, the KCL equation at the non-reference node is not an essential KCL equation, but you can use it to calculate the voltage source's current at a later time. For a voltage source that does not connect to the reference node, the KCL equation at either of it's nodes includes the current of the voltage source, so neither of these equations are essential KCL equations. However, the sum of these two KCL equations does not depend on the voltage-source current, because this current occurs as a positive term in one of the equations and as a negative term in the other. Adding the two equations gives an equation that does not include the voltage source current. If no more voltage sources connect to either of these nodes, this node pair is a *supernode.*

Generally, a supernode is a collection of nodes having node voltages that depend on only one node voltage and a set of voltage sources. The KCL equation of the supernode consists of the sum of currents having directions out of a surface enclosing the set of nodes. This supernode KCL equation equals the sum of the individual KCL equations for each node in the supernode set. The sum of KCL node equations forming the supernode equation eliminates the individual voltage source currents, because each voltage source current appears twice, first as a positive term and then as a negative term. Example 3.2 shows how to implement nodal analysis using these ideas.

Example 3.3 Circuit with Voltage Sources

For the circuit shown in Fig. 3.7

a) Select an independent set of nodes or supernodes. Define all node and supernode voltages. Draw the surface denoting the supernode.

b) Write the KCL equations for each node in the supernode. Sum these equations to obtain the supernode KCL equation. Examine this equation to see that it coincides with the KCL equation of the supernode.

c) Write the remaining KCL equation(s) for the circuit.

d) Rearrange the set of essential equations, collecting coefficients.

e) Express this set of equations as a single matrix equation.

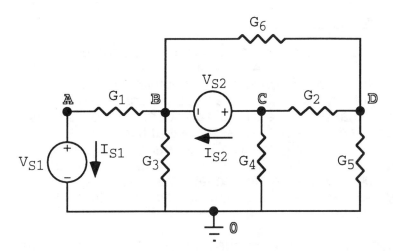

Figure 3.7 Circuit Example to Illustrate Nodal analysis with Voltage Sources.

Solution

a) The circuit has n = 4 and n_{vs} = 2, so the number of independent nodes is

$$\text{Number of KCL Equations} = n - n_{vs} = 4 - 2 = 2$$

The circuit of Fig. 3.7 is redrawn in Fig. 3.8, showing a selection of independent nodes, defining the node and supernode voltages, and showing the supernode surface. The label V_{S1} replacing node label A shows that voltage source V_{S1} sets this node voltage. We select the node voltage at node B to be node-voltage V_{N1}. Then, by KVL, the node voltage at node C equals $V_{N1}+V_{S2}$. If we choose the voltage at node C, instead of node B, to be node-voltage V_{N1} then the voltage at node B becomes $V_{N1} - V_{S2}$. The choice is arbitrary but binding for the rest of the analysis. Node D is the second independent node and has the label V_{N2}.

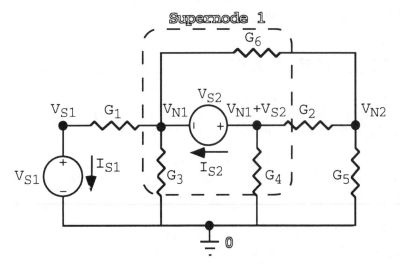

Figure 3.8 Circuit of Fig. 3.4 Showing Nodes and Supernodes.

b) The two KCL equations are

$$G_1\left(V_{N1} - V_{S1}\right) + G_3 V_{N1} + G_6\left(V_{N1} - V_{N2}\right) - I_{S2} = 0$$

$$G_2\left[\left(V_{N1} + V_{S2}\right) - V_{N2}\right] + G_4\left(V_{N1} + V_{S2}\right) + I_{S2} = 0$$

Adding these gives

$$G_1\left(V_{N1} - V_{S1}\right) + G_3 V_{N1} + G_6\left(V_{N1} - V_{N2}\right) + G_2\left[\left(V_{N1} + V_{S2}\right) - V_{N2}\right] + G_4\left(V_{N1} + V_{S2}\right) = 0$$

This equation is KCL for the supernode in Fig. 3.4.

c) The remaining essential KCL node equation at node 2 is

$$G_2\left[V_{N2} - \left(V_{N1} + V_{S2}\right)\right] + G_5 V_{N2} + G_6\left[V_{N2} - V_{N1}\right] = 0$$

d) Collecting terms for the supernode and node equations gives

$$\left(G_1 + G_2 + G_3 + G_4 + G_6\right)V_{N1} - \left(G_2 + G_6\right)V_{N2} = G_1 V_{S1} - \left(G_2 + G_4\right)V_{S2}$$

$$-\left(G_2 + G_6\right)V_{N1} + \left(G_2 + G_5 + G_6\right)V_{N2} = G_2 V_{S2}$$

e) The matrix form of these two KCL equations is

$$\begin{bmatrix} G_1 + G_2 + G_3 + G_4 + G_6 & -\left(G_2 + G_6\right) \\ -\left(G_2 + G_6\right) & G_2 + G_5 + G_6 \end{bmatrix}\begin{bmatrix} V_{N1} \\ V_{N2} \end{bmatrix} = \begin{bmatrix} G_1 V_{S1} - \left(G_2 + G_4\right)V_{S2} \\ G_2 V_{S2} \end{bmatrix}$$

$$= \begin{bmatrix} G_1 \\ 0 \end{bmatrix}V_{S1} + \begin{bmatrix} -\left(G_2 + G_4\right) \\ G_2 \end{bmatrix}V_{S2} = \begin{bmatrix} G_1 & \vdots & -\left(G_2 + G_4\right) \\ 0 & \vdots & G_2 \end{bmatrix}\begin{bmatrix} V_{S1} \\ V_{S2} \end{bmatrix}$$

In Example 3.3, conductances G_1, G_2, G_3, G_4, and G_6 are self conductances for supernode 1, because each conductance connects to the supernode. Conductances G_2, G_5, and G_6 connect to node 2, so are self conductances for node 2. Both G_2 and G_6 are mutual conductances between supernode 1 and node 2, so they accumulate negatively in the conductance matrix in row 1, column 2 and in row 2, column 1. Also, G_2 and G_6 connect between two non-reference nodes so both appear as a typical pattern-of-four, with two positive diagonal entries and two negative off-diagonal entries. The remaining conductances connect from one node to the reference node, so they appear only as additive terms in the appropriate diagonal positions.

Another useful viewpoint is to think of the current components due only to one source or node voltage at a time. For example, with V_{S1}, V_{S2}, and V_{N2} set to zero, the sum of currents away from the supernode is

$$\left(G_1 + G_2 + G_3 + G_4 + G_6\right)V_{N1}$$

This term accounts for the entry in the first row and first column of the conductance matrix. Current into node 2 due to V_{N1} alone is

$$(G_2 + G_6)V_{N1}$$

which accounts for the entry in row 1, column 2 of the conductance matrix.

Independent source terms appear in the right-hand side of the matrix equation. With V_{S1} alone not zero, there is a current G_1V_{S1} in conductance G_1 with direction towards supernode 1. This fact accounts for the nonzero entry in row 1 of the V_{S1} coefficient matrix. With V_{S2} alone having a nonzero value, conductance G_2 has a current component G_2V_{S2} with direction towards node 2 and away from supernode 1. Since a current with direction towards a node is positive on the right-hand-side of the KCL equations, this fact accounts for the positive G_2 entry in row 2 and the negative G_2 entry in row 1 of the V_{S2} coefficient matrix.

The point of these observations is that entry of every element in the matrix formulation of the essential KCL equations can be done by direct inspection of the circuit diagram. Writing the matrix form of the solution directly by inspection of the circuit saves time and reduces the possibility of making errors in writing and transcribing the KCL equations.

3.1.3 With VCCS Dependent Sources

Entry of a voltage-controlled current source (VCCS) into a circuit's conductance matrix is simple and straightforward. For example, a VCCS having a value g_m and connecting from node i to node j while under control of the node voltage of node k minus the node voltage at node 1 gives the following terms in the node-voltage KCL equations

$$
\begin{array}{ll}
\vdots & \\
i & \ldots + g_m\left(V_{Nk} - V_{Nl}\right) + \ldots = \ldots \\
\vdots & \\
j & \ldots - g_m\left(V_{Nk} - V_{Nl}\right) + \ldots = \ldots \\
\vdots &
\end{array}
\tag{3.3}
$$

You can see that a pattern of four g_m values occurs. The positive entries appear in row i and column k (the "from" row and the "+" controlling column) and in row j and column 1 (the "to" row and the "-" controlling column). The negative terms appear in the other square locations (the "to" row and the "+" controlling column and the "from" row and the "-" controlling column). This pattern is the same as for a conductance when i equals k and j also equals 1. This fact leads to the observation that a VCCS is the same as a conductance when the control nodes are the same as the nodes of the source. In general, a VCCS destroys the symmetry of the conductance matrix when i and k are not the same or j and 1 differ. If either i or j become the reference node, then that row falls outside of the matrix and only the two terms of the remaining row appear in the matrix. If either k or 1 become the reference node, then that column falls outside of the conductance matrix and only the two terms of the remaining column appear. If one row and one column correspond to the reference node, then only one term appears in the matrix. Of course having both rows or both columns become the reference node is possible, but is a trivial situation. The help documentation for a new MATLAB function gmAcc that accumulates a transconductance g_m onto a conductance matrix appears in Fig. 3.9.

```
amat=gmAcc(j,jp,k,kp,val,bmat)
   Accumulates gm value "val" onto matrix bmat.
   j and jp are + and - control nodes.
   k and kp are from and to nodes.
```

Figure 3.9 Function gmAcc Help.

Example 3.4 Circuit with VCCS Dependent Source

For the circuit shown in Fig. 3.10, write a MATLAB M-file to evaluate the conductance matrix G_{nn} and the current source matrix I_s, and solve for the node-voltage vector V_n.

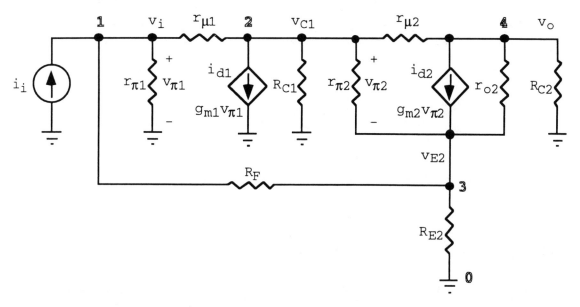

Figure 3.10 Transistor Feedback Circuit Example (g_{m1} = g_{m2} = 40 mS, $r_{\pi1}$ = $r_{\pi2}$ = 2.5 kΩ, r_{o2} = 100 kΩ, $r_{\mu1}$ = $r_{\mu2}$ = 10 MΩ, R_{C1} = 10 kΩ, R_{C2} = 8 kΩ, R_{E2} = 1.3 kΩ, R_F = 10 kΩ, i_i = 1 mA).

Solution

An M-file to solve this circuit appears in Fig. 3.11. The solution appears in Fig. 3.12. The units of the solution are V, mA, and kΩ for voltage, current, and resistance. This M-file defines the conductance values, then accumulates these onto the 4 × 4 conductance matrix G_{nn} using function nmAcc. Function gmAcc accumulates the two transconductance sources onto G_{nn}. Finally, function srcAcc enters the 1-mA current source I_i into the 4×1 source matrix I_s. The computation shows that the transconductance gain of this feedback amplifier is 67.7238 V/mA

```
% Example 3.4 - Circuit with VCCS Dependent Source
clear
gm=40;
gpi=1/2.5;   go=1/100;    gmu=1/1E4;
GC1=1/10;    GC2=1/8;     GE2=1/1.3;   GF=1/10;
Gnn=nmAcc(1,0,gpi,zeros(4));
Gnn=nmAcc(1,2,gmu,Gnn);
Gnn=nmAcc(2,0,GC1,Gnn);;
Gnn=nmAcc(2,3,gpi,Gnn);
Gnn=nmAcc(2,4,gmu,Gnn);
Gnn=nmAcc(4,3,go,Gnn);
Gnn=nmAcc(4,0,GC2,Gnn);
Gnn=nmAcc(3,0,GE2,Gnn);
Gnn=nmAcc(1,3,GF,Gnn);
Gnn=gmAcc(1,0,2,0,gm,Gnn);
Gnn=gmAcc(2,3,4,3,gm,Gnn)
Is=srcAcc(0,1,1,zeros(4,1));
Vn=Gnn\Is
```

Figure 3.11 M-file to Solve Example 3.4.

```
Gnn  =
     0.5001    -0.0001    -0.1000          0
    39.9999     0.5002    -0.4000    -0.0001
    -0.1000   -40.4000    41.2792    -0.0100
          0    39.9999   -40.0100     0.1351
Vn  =
     0.0277
   -10.0824
    -9.8512
    67.7238
```

Figure 3.12 Solution to Example 3.4.

3.1.4 Nodal Equations with Dependent Sources[1]

A general form of the nodal analysis KCL matrix equation is

$$G_{nn}V_n + A_{nd}X_d = A_{ns}X_s \qquad (3.4)$$

where V_n is the node-voltage solution vector, X_d is a vector of dependent sources, and X_s is a vector of independent sources. To have conductance matrix G_{nn} show symmetry, write the KCL node equations in the same order as the order of the node voltage components in the solution vector, add currents away from each node, and include VCCS sources in A_{nd}. However, you can embed VCCS sources in G_{nn} as in Example 3.4 if you wish. Some of the KCL equations in Eq. (3.4) may be supernode equations. Matrix G_{nn} accounts for current terms in the KCL node equations due to circuit conductances. Matrix

---

[1] A portion of this Section and Section 3.2.3 are reproduced in part from J. G. Gottling, "Node and mesh analysis by inspection," to be published in IEEE Transactions on Education, Nov., 1994, © 1994 IEEE.

$\mathcal{A}_{nd}$ accounts for contributions to the KCL equations due to dependent sources, and matrix $\mathcal{A}_{ns}$ inserts current components due to the source vector $\mathcal{X}_s$.

Assuming a linear dependence of the vector of dependent sources $\mathcal{X}_d$ on the vector of controlling variables $\mathcal{X}_c$, then

$$X_{d} = \mathcal{A}_{dc} X_{c} \tag{3.5}$$

Usually the controlled-source matrix $\mathcal{A}_{dc}$ is both square and diagonal, but this condition is not necessary. A dependent source can depend on two or more controlling variables or two or more dependent sources can have the same controlling variable.

Finally, assume that the controlling variable vector $\mathcal{X}_c$ depends linearly on the dependent-source vector, the node-voltage vector, and the independent-source vector, so that

$$X_{c} = \mathcal{A}_{cd} X_{d} + \mathcal{A}_{cn} V_{n} + \mathcal{A}_{cs} X_{s} \tag{3.6}$$

The individual equations in Eq. 3.6 are KVL or KCL equations that identify the controlling variables in terms of the various node voltages and dependent or independent sources.

Substitution of Eq. 3.6 into Eq. 3.5 and solving for $\mathcal{X}_d$ gives

$$X_{d} = \left(I_{d} - \mathcal{A}_{dc}\mathcal{A}_{cd} \right)^{-1} \mathcal{A}_{dc} \left(\mathcal{A}_{cn} V_{n} + \mathcal{A}_{cs} X_{s} \right) \tag{3.7}$$

where I_d is the identity matrix with order equal to the number of dependent sources in $\mathcal{X}_d$. Alternately, substitution of Eq. 3.5 into Eq. 3.6, solving for $\mathcal{X}_c$, and substitution back into Eq. 3.4 gives

$$X_{d} = \mathcal{A}_{dc} \left(I_{c} - \mathcal{A}_{cd}\mathcal{A}_{dc} \right)^{-1} \left(\mathcal{A}_{cn} V_{n} + \mathcal{A}_{cs} X_{s} \right) \tag{3.8}$$

where I_c is an identity matrix with order equal to the number of controlling variables in $\mathcal{X}_c$. Equations Eq. 3.7 and Eq. 3.8 are equivalent, since

$$\left(I_{d} - \mathcal{A}_{dc}\mathcal{A}_{cd} \right)^{-1} \mathcal{A}_{dc} = \mathcal{A}_{dc} \left(I_{c} - \mathcal{A}_{cd}\mathcal{A}_{dc} \right)^{-1} \tag{3.9}$$

Finally, substitution of Eq. 3.7 into Eq. 3.4 and collecting coefficients gives the general node voltage equations

$$\left\{ G_{nn} + \mathcal{A}_{nd} \left(I_{d} - \mathcal{A}_{dc}\mathcal{A}_{cd} \right)^{-1} \mathcal{A}_{dc}\mathcal{A}_{cn} \right\} V_{n} = \left\{ \mathcal{A}_{ns} - \mathcal{A}_{nd} \left(I_{d} - \mathcal{A}_{dc}\mathcal{A}_{cd} \right)^{-1} \mathcal{A}_{dc}\mathcal{A}_{cs} \right\} X_{s} \tag{3.10}$$

If the circuit contains only one dependent source then the matrix factor $I_d - \mathcal{A}_{dc}\mathcal{A}_{cd}$ becomes a scalar and is known in feedback theory as the return difference. We call this quantity the return difference matrix. The return difference matrix plays a less significant role here than in feedback theory, because $\mathcal{A}_{cd}$ is null whenever the controlling variable is not directly dependent on any of the dependent sources. Node-voltage solution variables often intervene between the dependent sources and their controlling variables. If a circuit has one or more dependent sources, then the inverse of the return difference matrix has to exist for the circuit to have a solution.

The node voltages may not be the solution variables of interest. Output variables as a group constitute an output vector $\mathcal{X}_o$. The output vector is a linear function of the dependent source vector X_d, the node-voltage vector V_n, and the source vector X_s, so

$$X_o = A_{od}X_d + A_{on}V_n + A_{os}X_s \tag{3.11}$$

Substitution of Eq. 3.7 into Eq. 3.11 gives

$$X_o = \left[A_{on} + A_{od}(I - A_{dc}A_{cd})^{-1}A_{dc}A_{cn}\right]V_n + \left[A_{os} + A_{od}(I - A_{dc}A_{cd})^{-1}A_{dc}A_{cs}\right]X_s \tag{3.12}$$

which expresses the output vector in terms of the node-voltage vector and the independent-source vector.

Another way to solve Eqs. 3.4, 3.5, 3.6, and 3.11 is to rewrite them as

$$G_{nn}V_n + A_{nd}X_d + 0X_c + 0X_o = A_{ns}X_s \tag{3.13}$$

$$0V_n + IX_d - A_{dc}X_c + 0X_o = 0 \tag{3.14}$$

$$-A_{cn}V_n - A_{cd}X_d + IX_c + 0X_o = A_{cs}X_s \tag{3.15}$$

$$-A_{on}V_n - A_{od}X_d + 0X_c + IX_o = A_{os}X_s \tag{3.16}$$

Now, write these as one matrix equation

$$\begin{bmatrix} G_{nn} & A_{nd} & 0 & 0 \\ 0 & I & -A_{dc} & 0 \\ -A_{cn} & -A_{cd} & I & 0 \\ -A_{on} & -A_{od} & 0 & I \end{bmatrix} \begin{bmatrix} V_n \\ X_d \\ X_c \\ X_o \end{bmatrix} = \begin{bmatrix} A_{ns} \\ 0 \\ A_{cs} \\ A_{os} \end{bmatrix} X_s \tag{3.17}$$

This super-matrix equation displays the circuit parameters in simple form, avoids the complexity of the solution Eq. 3.12, and allows use of the adjoint system method[2].

Example 3.5 Example with CCVS and CCCS

For the circuit shown in Fig. 3.13, which has both a CCVS and a CCCS source

a) Prepare the circuit for nodal analysis by inspection. Mark the nodes and supernodes with appropriate node voltages and indicate any supernode surfaces.

b) Write the essential KCL node or supernode equations in terms of the node voltages, dependent sources, and independent source. Rearrange these, collect coefficients, and move the independent source term(s) to the right-hand side. Write these in matrix form (See Eq. 3.4).

c) Write the dependency relations for I_{D1} and V_{D2} in the form of Eq. 3.5.

d) Write the relations for controlling variables I_{R1} and I_{R3} in terms of the dependent sources, node voltages, and independent sources. Express

[2] J. Vlach & K. Singhal, *Computer Methods for Circuit Analysis and Design*, Van Nostrand Reinhold, New York, 1983, pp. 171-261.

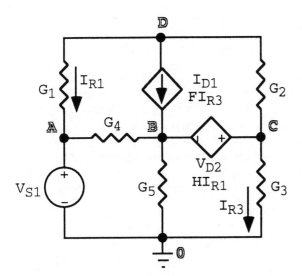

Figure 3.13 Circuit for Example 3.5 with CCVS and CCCS Elements (V_{S1} = 10 V, $G_1 = G_2 = G_3 = G_4 = G_5 = 2$ S, F = 10, and H = 5 Ω).

these in the matrix form Eq. 3.6. Examine the matrix form and convince yourself that you can write this equation by inspection of the circuit.

e) Use MATLAB to calculate the node-voltage vector using both the matrix solution, Eq. 3.10, and the super-matrix method, Eq. 3.17.

Solution

a) Figure 3.14 shows the circuit diagram with node voltages and supernode surfaces. For this circuit n+1 equals 5 and n_{vs} equals 2, so the number of independent node voltages is

$$\text{Number of KCL Equations } = n - n_{vs} = 4 - 2 = 2$$

The circuit diagram shows that we choose the middle node to be node 1 and the top node is node 2. Source v_{S1} gives the node voltage at node A, and dependent source v_{D2} defines the voltage of node C relative to node B, which is node 1. Nodes B and C are the supernode 1.

b) The supernode 1 and node 2 equations are

$$G_2\left[\left(V_{N1} + V_{D2}\right) - V_{N2}\right] + G_3\left(V_{N1} + V_{D2}\right) + G_4\left(V_{N1} - V_{S1}\right) + G_5 V_{N1} - I_{D1} = 0$$

$$G_1\left(V_{N2} - V_{S1}\right) + G_2\left[V_{N2} - \left(V_{N1} + V_{D2}\right)\right] + I_{D1} = 0$$

Collecting coefficients

$$\left(G_2 + G_3 + G_4 + G_5\right)V_{N1} - G_2 V_{N2} - I_{D1} + \left(G_2 + G_3\right)V_{D2} = G_4 V_{S1}$$

$$-G_2 V_{N1} + \left(G_1 + G_2\right)V_{N2} + I_{D1} - G_2 V_{D2} = G_1 V_{S1}$$

In the matrix form of Eq. 3.4, these become

$$\begin{bmatrix} G_2 + G_3 + G_4 + G_5 & -G_2 \\ -G_2 & G_1 + G_2 \end{bmatrix}\begin{bmatrix} V_{N1} \\ V_{N2} \end{bmatrix} + \begin{bmatrix} -1 & G_2 + G_3 \\ 1 & -G_2 \end{bmatrix}\begin{bmatrix} I_{D1} \\ V_{D2} \end{bmatrix} = \begin{bmatrix} G_4 \\ G_1 \end{bmatrix} V_{S1}$$

Each matrix coefficient is evident by inspection of the circuit. In particular G_2, G_3, G_4, and G_5 are self-conductance terms for supernode 1 and G_1 and G_2 are self-conductance terms for node 2. G_2 appears in the typical pattern of four because it connects between supernode 1 and node 2. The entries in the first column of the $\mathcal{A}_{nd}$ matrix are the ratios of current components at supernode 1 and node 2 to the controlled current i_{D1}. The second column values give the ratios of currents away from supernode 1 and node 2 to the controlled voltage V_{D2}. The entries in $\mathcal{A}_{ns}$ are the ratios of the current components into the supernode and node due to independent source v_{S1}.

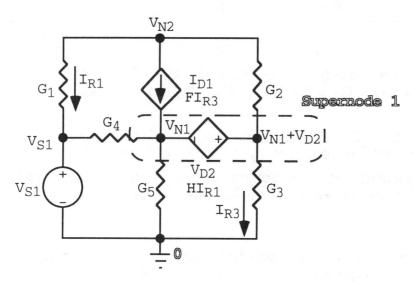

Figure 3.14 Circuit for Example 3.5 after Preparation for Nodal Analysis.

c) For this circuit Eq. 3.5 is

$$\begin{bmatrix} I_{D1} \\ V_{D2} \end{bmatrix} = \begin{bmatrix} F & 0 \\ 0 & H \end{bmatrix}\begin{bmatrix} I_{R3} \\ I_{R1} \end{bmatrix}$$

d) The controlling variables I_{R1} and I_{R3} are

$$I_{R3} = G_3\left(V_{N1} + V_{D2}\right)$$

$$I_{R1} = G_1\left(V_{N2} - V_{S1}\right)$$

In the form of Eq. 3.6 these are

$$\begin{bmatrix} I_{R3} \\ I_{R1} \end{bmatrix} = \begin{bmatrix} 0 & G_3 \\ 0 & 0 \end{bmatrix}\begin{bmatrix} I_{D1} \\ V_{D2} \end{bmatrix} + \begin{bmatrix} G_3 & 0 \\ 0 & G_1 \end{bmatrix}\begin{bmatrix} V_{N1} \\ V_{N2} \end{bmatrix} + \begin{bmatrix} 0 \\ -G_1 \end{bmatrix} V_{S1}$$

The component values for the coefficient matrices above are evident by inspection of the circuit. The first-row terms account for Ohm's law for the I_{R1} current, and the second-row terms are Ohm's law terms for the I_{R3} current.

e) The MATLAB M-file shown in Fig. 3.15 gives the numerical solution for this problem that you see in Fig. 3.16. Documentation for function genAnal, which finds the nodal analysis solution using Eq. 3.10, is in Fig. 3.17. The super matrix solution, using Eq. 3.17, appears in the last part of the M-file in Fig. 3.15. The first two components of the super matrix solution vector are the node voltages. The next two are the values of the dependent sources I_{D1} and V_{D2}. The last two values are the controlling currents i_{R3} and i_{R1}.

```
% Example 3.5 - Example with CCVS and CCCS
clear
format compact
G1=2; G2=2; G3=2; G4=2; G5=2;
% Define the conductance matrix.
Gnn=nmAcc(1,2,G2,zeros(2,2));
Gnn=nmAcc(1,0,G3+G4+G5,Gnn);
Gnn=nmAcc(2,0,G1,Gnn);
% Define And
And=[-1 G2+G3;1 -G2];
% Define Ans
Ans=[G4;G1];
% Define Adc
Adc=[10 0;0 5];
% Define Acd
Acd=[0 G3;0 0];
% Define Acn
Acn=[G3 0;0 G1];
% Define Acs
Acs=[0;-G1];
% Define Source Vector
Vs1=10;
disp('Node-voltage vector Vn using genAnal:')
Vn=genAnal(Vs1,Gnn,And,Ans,Adc,Acd,Acn,Acs);
disp(Vn)
disp('Super matrix Gsuper:')
[rGnn,cGnn]=size(Gnn);
[rAdc,cAdc]=size(Adc);
[rAcd,cAcd]=size(Acd);
Znc=zeros(rGnn,cAdc);
Zdn=zeros(rAdc,cGnn);
Ud=eye(rAdc);
Uc=eye(rAcd);
Gsuper=[Gnn And Znc;Zdn Ud -Adc;-Acn -Acd Uc];
disp(Gsuper)
XSuper=[Ans;zeros(rAdc,1);Acs];
disp('Vn, Xd, and Xc')
Ysuper=Gsuper\XSuper*Vs1;
disp(Ysuper)
```

Figure 3.15 MATLAB M-File Solution for Example 3.5.

```
»Node-voltage vector Vn using genAnal:
    5.8192
    9.3220
Super matrix Gsuper:
        8       -2      -1       4       0       0
       -2        4       1      -2       0       0
        0        0       1       0     -10       0
        0        0       0       1       0      -5
       -2        0       0      -2       1       0
        0       -2       0       0       0       1
Vn,  Xd,  and Xc
    5.8192
    9.3220
  -19.2090
   -6.7797
   -1.9209
   -1.3559
```

Figure 3.16 MATLAB Solution of Example 3.5 using Eq. 3.9.

```
xx=genAnal(xs,axx,axd,axs,adc,acd,acx,acs)
  Calculates the node-voltage or mesh-current vector
  xs  - Source vector
  axx - Node or mesh matrix
  axd - Dependent source matrix
  adc - Relates dependent sources to controlling variables
  acd - Relates controlling variables to dependent sources
  acx - Relates controlling variables to mesh voltages or
        mesh currents
  acs - Relates controlling variables to independent sources
```

Figure 3.17 genAnal Help.

3.1.5 Op-Amp Nodal Analysis

Operational amplifier (op amp) circuit analysis usually involves nodal analysis. Solving circuits containing op amps using matrix analysis is not difficult. However the G_{nn} matrix is not symmetric. Write the matrix node equation, but recognize that the matrix rows correspond to nodal KCL equations, while the corresponding columns may relate to different node voltages. Do not write KCL equations at the output terminal of any op amp, because that KCL equation has to include an op amp output current. Analysis of the two op-amp circuit in Fig. 3.18, illustrates the procedure.

Example 3.6 Op-Amp Nodal Analysis Example

For the circuit shown in Fig. 3.18, write the nodal analysis equations.

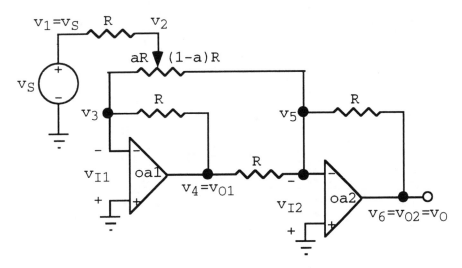

Figure 3.18 Op-Amp Circuit Example.

Solution

$$
\begin{bmatrix}
\dfrac{1}{R}+\dfrac{1}{aR}+\dfrac{1}{(1-a)R} & 0 & 0 \\[2ex]
-\dfrac{1}{aR} & -\dfrac{1}{R} & 0 \\[2ex]
-\dfrac{1}{(1-a)R} & -\dfrac{1}{R} & -\dfrac{1}{R}
\end{bmatrix}
\begin{bmatrix} v_2 \\ v_4 \\ v_o \end{bmatrix}
=
\begin{bmatrix} 1/R \\ 0 \\ 0 \end{bmatrix} v_S
$$

In the nodal analysis matrix equation, the first row elements account for terms in the KCL equation at node 2. The second and third rows account for terms in the KCL equations at the inverting terminals of the op amps. If the circuit has controlled sources in addition to the op amps, then write this matrix equation in the form of Eqs. 3.4 through 3.6 to account for these and solve using Eq. 3.10 or Eq. 3.17.

The above method follows from the general method of Eqs. 3.4 through 3.6. Write Eq. 3.4, resolving node voltage vector $\mathcal{V}_n$ into four components and dependent-source vector $\mathcal{X}_d$ into two components

$$
\begin{bmatrix} \mathcal{G}_{nn1} & \vdots & \mathcal{G}_{n2} & \vdots & \mathcal{G}_{n3} & \vdots & \mathcal{G}_{n4} \end{bmatrix}
\begin{bmatrix} \mathcal{V}_{n1} \\ \overline{\mathcal{V}_{n2}} \\ \overline{\mathcal{V}_{n3}} \\ \overline{\mathcal{V}_{n4}} \end{bmatrix}
+
\begin{bmatrix} \mathcal{A}_{nd1} & \vdots & \mathcal{A}_{nd2} \end{bmatrix}
\begin{bmatrix} \mathcal{X}_{d1} \\ \overline{\mathcal{X}_{d2}} \end{bmatrix}
= \mathcal{A}_{ns}\mathcal{X}_s
\qquad (3.18)
$$

Here $\mathcal{V}_{n4}$ contains all op-amp non-ground input node voltages for op amps having their other input terminal at ground. Node-voltage partitions $\mathcal{V}_{n2}$ and $\mathcal{V}_{n3}$ include the inverting and noninverting op-amp input node voltages for op amps not having either input at ground. Partition $\mathcal{V}_{n1}$ has all of the remaining node voltages. Partition $\mathcal{X}_{d1}$ accounts for currents due to controlled sources. The $\mathcal{X}_{d2}$ term accounts for all current terms due to the op amps, which are modeled as voltage-controlled voltage sources. The controlling relations Eq. 3.5 take the form

$$X_d = \begin{bmatrix} X_{d1} \\ \hline X_{d2} \end{bmatrix} = \begin{bmatrix} \mathcal{A}_{dc11} & \vdots & 0 \\ \hline 0 & \vdots & A_{dc22} \end{bmatrix} \begin{bmatrix} X_{c1} \\ \hline X_{c2} \end{bmatrix} = \mathcal{A}_{dc} X_c \qquad (3.19)$$

and the controlling variable equation Eq. 3.6 is

$$X_c = \begin{bmatrix} X_{c1} \\ \hline X_{c2} \end{bmatrix} = \begin{bmatrix} \mathcal{A}_{cd11} & \vdots & \mathcal{A}_{cd12} \\ \hline \mathcal{A}_{cd21} & \vdots & A_{cd22} \end{bmatrix} \begin{bmatrix} X_{d1} \\ \hline X_{d2} \end{bmatrix} + \begin{bmatrix} \mathcal{A}_{cn1} & \vdots & \mathcal{A}_{cn2} & \vdots & \mathcal{A}_{cn3} & \vdots & \mathcal{A}_{cn4} \end{bmatrix} \begin{bmatrix} V_{n1} \\ \hline V_{n2} \\ \hline V_{n3} \\ \hline V_{n4} \end{bmatrix} + \mathcal{A}_{cs} X_s$$

$$(3.20)$$

In Eq. 3.20 the relationship between op-amp vectors X_{d2} and X_{c2} assures that all components of X_{c2} go to zero as the op-amp voltage gains become infinite. The set of equations in Eq. 3.20 that involve X_{c2} assures that $V_{n2} = V_{n3}$ and $V_{n4} = 0$. Then Eq. 3.18 becomes

$$\begin{bmatrix} \mathcal{G}_{nn1} & \vdots & \mathcal{G}_{nn2} + \mathcal{G}_{nn3} \end{bmatrix} \begin{bmatrix} V_{n1} \\ V_{n2} \end{bmatrix} + \mathcal{A}_{nd1} X_{d1} + \mathcal{A}_{nd2} X_{d2} = \mathcal{A}_{ns} X_s \qquad (3.21)$$

The non op-amp equations in Eqs. 3.19 and 3.20 are

$$X_{d1} = \mathcal{A}_{dc11} X_{c1} \qquad (3.22)$$

$$X_{c1} = \mathcal{A}_{cd11} X_{d1} + \mathcal{A}_{cd12} X_{d2} + \begin{bmatrix} \mathcal{A}_{cn1} & \vdots & \mathcal{A}_{cn2} + \mathcal{A}_{cn3} \end{bmatrix} \begin{bmatrix} V_{n1} \\ V_{n2} \end{bmatrix} + \mathcal{A}_{cs} X_s \qquad (3.23)$$

Solving for X_{d1}

$$X_{d1} = \left(1 - \mathcal{A}_{dc11} \mathcal{A}_{cd11}\right)^{-1} \mathcal{A}_{dc11} \left\{ \mathcal{A}_{cd12} X_{d2} + \begin{bmatrix} \mathcal{A}_{cn1} & \vdots & \mathcal{A}_{cn2} + \mathcal{A}_{cn3} \end{bmatrix} \begin{bmatrix} V_{n1} \\ V_{n2} \end{bmatrix} + \mathcal{A}_{cs} X_s \right\}$$

$$(3.24)$$

and substituting into Eq. 3.18 gives the general solution

$$\begin{bmatrix} \mathcal{G}_{nn1} & \vdots & \mathcal{G}_{nn2} + \mathcal{G}_{nn3} & \vdots & \mathcal{A}_{nd2} + \mathcal{A}_{nd1}\left(1 - \mathcal{A}_{dc11} \mathcal{A}_{cd11}\right)^{-1} \mathcal{A}_{dc11} \mathcal{A}_{cd12} \end{bmatrix} \begin{bmatrix} V_{n1} \\ \hline V_{n2} \\ \hline X_{d2} \end{bmatrix}$$

$$= \left\{ \mathcal{A}_{ns} - \mathcal{A}_{nd1}\left(1 - \mathcal{A}_{dc11} \mathcal{A}_{cd11}\right)^{-1} \mathcal{A}_{dc11} \mathcal{A}_{cs} \right\} X_s \qquad (3.25)$$

When the partition given above is not possible, the procedure changes but still involves a set of equations similar to 3.19 to 3.23. First add columns in matrices $\mathcal{G}_{nn}$ and $\mathcal{A}_{cs}$ that correspond to rows of V_n for node voltages that op amps constrain to be the same. Next, replace the first of these columns with this sum and delete the remaining columns in this set. Delete all but the first row in V_n of the corresponding node voltages. Finally solve Eq. 3.22 and Eq. 3.23 for X_{d1} and obtain Eq. 3.25 as above.

For the circuit of Figure 3.19, the general procedure involving Eq. 3.18 through Eq. 3.25 gives

$$
\begin{bmatrix}
\dfrac{1}{R}+\dfrac{1}{aR}+\dfrac{1}{(1-a)R} & -\dfrac{1}{aR} & -\dfrac{1}{(1-a)R} \\
-\dfrac{1}{aR} & \dfrac{1}{R}+\dfrac{1}{aR} & 0 \\
-\dfrac{1}{(1-a)R} & 0 & \dfrac{2}{R}+\dfrac{1}{(1-a)R}
\end{bmatrix}
\begin{bmatrix} v_2 \\ v_3 \\ v_5 \end{bmatrix}
+
\begin{bmatrix}
0 & 0 \\
-\dfrac{1}{R} & 0 \\
-\dfrac{1}{R} & -\dfrac{1}{R}
\end{bmatrix}
\begin{bmatrix} v_{O1} \\ v_{O2} \end{bmatrix}
$$

$$
=\begin{bmatrix} \dfrac{1}{R} \\ 0 \\ 0 \end{bmatrix} v_S
\tag{3.26}
$$

$$
X_{d2}=\begin{bmatrix} v_{O1} \\ v_{O2} \end{bmatrix}=\begin{bmatrix} A_1 & 0 \\ 0 & A_2 \end{bmatrix}\begin{bmatrix} v_{I1} \\ v_{I2} \end{bmatrix}
\tag{3.27}
$$

$$
X_{c2}=\begin{bmatrix} v_{I1} \\ v_{I2} \end{bmatrix}=\begin{bmatrix} 0 & 0 \\ 0 & 0 \end{bmatrix}\begin{bmatrix} v_{O1} \\ v_{O2} \end{bmatrix}+\begin{bmatrix} 0 & -1 & 0 \\ 0 & 0 & -1 \end{bmatrix}\begin{bmatrix} v_2 \\ v_3 \\ v_5 \end{bmatrix}+\begin{bmatrix} 0 \\ 0 \end{bmatrix}v_S
\tag{3.28}
$$

In the limit as A_1 and A_2 go to infinity, v_{I1} and v_{I2} go to zero. Then Eq. 3.28 asserts that v_3 and v_5 are zero. Using these conclusions in Eq. 3.26 gives

$$
\begin{bmatrix}
\dfrac{1}{R}+\dfrac{1}{aR}+\dfrac{1}{(1-a)R} & 0 & 0 \\
-\dfrac{1}{aR} & -\dfrac{1}{R} & 0 \\
-\dfrac{1}{(1-a)R} & \dfrac{1}{R} & -\dfrac{1}{R}
\end{bmatrix}
\begin{bmatrix} v_2 \\ v_{O1} \\ v_{O2} \end{bmatrix}
=\begin{bmatrix} 1/R \\ 0 \\ 0 \end{bmatrix} v_S
\tag{3.29}
$$

as found before.

3.2 Mesh Analysis

Mesh analysis expresses a planar circuit's solution using a linearly-independent set of mesh currents. A mesh current is a current component that circulates through all elements on the perimeter of a window of a circuit's schematic diagram. By convention all mesh currents circulate in the same direction, usually clockwise. A mesh current physically is measurable if a circuit element's current has only one mesh current component. The circuit solution consists of a linearly-independent set of Kirchhoff voltage law (KVL) mesh current equations. We assume that the circuit is proper.

3.2.1 Mesh Analysis with Independent Voltage Sources

Mesh current analysis of a circuit having only independent voltage sources involves identifying a set of mesh currents and writing a number of independent KVL equations given by

$$\text{No. of Independent KVL Equations} = b - n \qquad (3.30)$$

You can write the independent set of equations in matrix form directly by inspection of the circuit. This fact arises directly from the summation form of KVL. Each resistance in the circuit has a current equal to the difference between two mesh currents and has a voltage proportional to both the mesh-current difference and the resistance value. If the sequence of the mesh currents in the mesh-current solution vector is the same as for the KVL equations, then the mesh equations have diagonal symmetry. For example, if a resistance R_A exists between mesh i and mesh j then the terms shown in Eq. 3.31 appear in the KVL equations

$$
\begin{aligned}
&\vdots \\
i \qquad &\ldots + R_A\left(I_{M_i} - I_{M_j}\right) + \ldots = \ldots \\
&\vdots \\
j \qquad &\ldots - R_A\left(I_{M_i} - I_{M_j}\right) + \ldots = \ldots \\
&\vdots
\end{aligned}
\qquad (3.31)
$$

The R_A term appears as a positive term in the KVL equations in the i-th and j-th diagonal positions and as a negative term in the i-th row and j-th column and in the j-th row and i-th column. If a resistance exists at the outer perimeter of the circuit diagram, then that resistance appears only as an additive term in the i-th diagonal position.

Example 3.7 Independent Voltage Source Mesh Analysis Example

For the circuit shown in Fig. 3.19

a) write the KVL mesh-current equations in terms of the individual resistance voltages and mesh currents I_{M1}, I_{M2}, and I_{M3}.

b) Rewrite the equations collecting coefficients common to each mesh current.

c) Express the three KVL mesh-current equations in matrix form. Notice the diagonal symmetry of the resistance matrix and the location of nonzero entries in the voltage source matrix.

d) Write the resistance matrix as the sum of five separate matrices, each involving only one resistance and the voltage source matrix as the sum of two matrices each involving only one of the voltage sources.

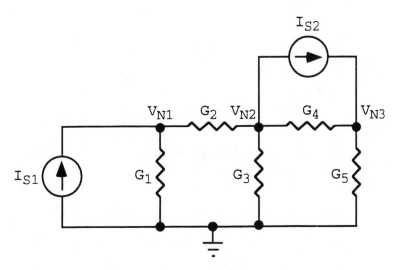

Figure 3.19 Circuit for Example 3.7.

For the circuit shown in Fig. 3.19

a) write the KVL mesh-current equations in terms of the individual resistance voltages and mesh currents I_{M1}, I_{M2}, and I_{M3}.

b) Rewrite the equations collecting coefficients common to each mesh current.

c) Express the three KVL mesh-current equations in matrix form. Notice the diagonal symmetry of the resistance matrix and the location of nonzero entries in the voltage source matrix.

d) Write the resistance matrix as the sum of five separate matrices, each involving only one resistance and the voltage source matrix as the sum of two matrices each involving only one of the voltage sources.

Solution

a) The three KVL equations are

$$R_1 I_{M1} + R_2 (I_{M1} - I_{M2}) = V_{S1}$$

$$R_2 (I_{M2} - I_{M1}) + R_3 I_{M2} + R_4 (I_{M2} - I_{M3}) = -V_{S2}$$

$$R_4 (I_{M3} - I_{M2}) + R_5 I_{M3} = V_{S2}$$

b) Collecting coefficients of each mesh current

$$(R_1 + R_2) I_{M1} - R_2 I_{M2} = V_{S1}$$

$$-R_2 I_{M1} + (R_2 + R_3 + R_4) I_{M2} - R_4 I_{M3} = -V_{S2}$$

$$-R_4 I_{M2} + (R_4 + R_5) I_{M3} = V_{S2}$$

c) Writing in matrix form

$$
\begin{bmatrix} R_1 + R_2 & -R_2 & 0 \\ -R_2 & R_2 + R_3 + R_4 & -R_4 \\ 0 & -R_4 & R_4 + R_5 \end{bmatrix} \begin{bmatrix} I_{M1} \\ I_{M2} \\ I_{M3} \end{bmatrix} = \begin{bmatrix} V_{S1} \\ -V_{S2} \\ V_{S2} \end{bmatrix}
$$

You can write this matrix solution directly by inspection of the circuit. Each term on the diagonal consists of the sum of resistances that each mesh current passes through. These sums are *self resistance* terms. The off-diagonal terms, known as *mutual resistance* terms, are the negative sum of resistances that one mesh shares with an adjacent mesh.

d) Rewriting the resistance and voltage-source matrices into separate terms

$$
\left\{ \begin{bmatrix} R_1 & 0 & 0 \\ 0 & 0 & 0 \\ 0 & 0 & 0 \end{bmatrix} + \begin{bmatrix} R_2 & -R_2 & 0 \\ -R_2 & R_2 & 0 \\ 0 & 0 & 0 \end{bmatrix} + \begin{bmatrix} 0 & 0 & 0 \\ 0 & R_3 & 0 \\ 0 & 0 & 0 \end{bmatrix} + \begin{bmatrix} 0 & 0 & 0 \\ 0 & R_4 & -R_4 \\ 0 & -R_4 & R_4 \end{bmatrix} \right.
$$

$$
\left. + \begin{bmatrix} 0 & 0 & 0 \\ 0 & 0 & 0 \\ 0 & 0 & R_5 \end{bmatrix} \right\} \begin{bmatrix} I_{M1} \\ I_{M2} \\ I_{M3} \end{bmatrix} = \left\{ \begin{bmatrix} V_{S1} \\ 0 \\ 0 \end{bmatrix} + \begin{bmatrix} 0 \\ -V_{S2} \\ V_{S2} \end{bmatrix} \right\}
$$

This matrix form of the KVL mesh equation shows how each circuit element contributes to the structure of the matrix equation solution. A resistance that exists on the outside perimeter of the circuit appears only in the diagonal position for that mesh. A resistance that exists between two meshes appears in four positions. Two of these on the diagonal are positive, and the remaining two off-diagonal terms are negative. The off-diagonal entries occur in the row of one mesh and the column of the other mesh.

To account for a voltage source, enter the voltage-source value in one or two rows of the right-hand side excitation matrix. The entry is positive in the row of a mesh if the voltage source is a rise. The entry is negative in the row of a mesh where the voltage source is a drop.

3.2.2 Mesh Analysis with Supermeshes

We have seen above how to handle circuits having independent or dependent voltage sources that constrain the difference between two node voltages in nodal analysis. Now, we must learn what to do in mesh analysis when a circuit has current sources that straddle two meshes. The following example shows what to do.

Example 3.8 Supermesh   Example

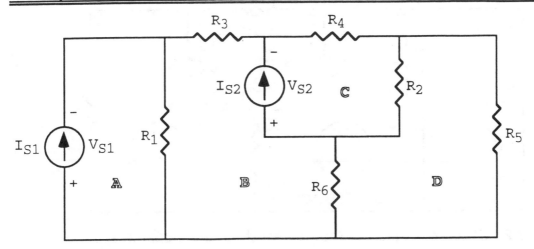

Figure 3.20 Circuit Example to Illustrate Mesh Analysis with Current Sources.

For the circuit shown in Fig. 3.20

a) Select an independent set of meshes or supermeshes. Define all mesh and supermesh currents. Draw the path denoting the supermesh.

b) Write the KVL equations for each mesh in the supermesh. Sum these equations to obtain the supermesh KVL equation. Examine this equation to see that it coincides with the KVL equation of the supermesh.

c) Write the remaining KVL equation(s) for the circuit.

d) Rearrange the set of essential equations, collecting coefficients.

e) Express this set of equations as a single matrix equation.

f) Use the new MATLAB functions nmAcc and srcAcc to build the resistance and excitation matrices when all resistances are 1 Ω. Find the mesh-vector solutions when I_{S1} = 1 A and I_{S2} = 0 A and when I_{S1} = 0 A and I_{S2} = 1 A.

Solution

a) The circuit has b = 8, n = 4 and n_{is} = 2, so the number of independent meshes is

$$\text{Number of KVL Equations} = b - n - n_{vs} = 8 - 4 - 2 = 2$$

The circuit of Fig. 3.20 is redrawn in Fig. 3.21, showing a selection of meshes, defining all mesh and supermesh currents and the supermesh path. The label I_{S1} at mesh A shows that the first source sets this mesh current. Mesh B becomes mesh 1 and has mesh current I_{M1}. By KCL, the mesh current at mesh C equals $I_{M1}+I_{S1}$. Choosing mesh B instead of mesh C to have mesh current I_{M1} is a convenient but arbitrary choice that makes the second mesh current in the supernode set depend on the first additively. Mesh D is the second independent mesh, so has the label 2.

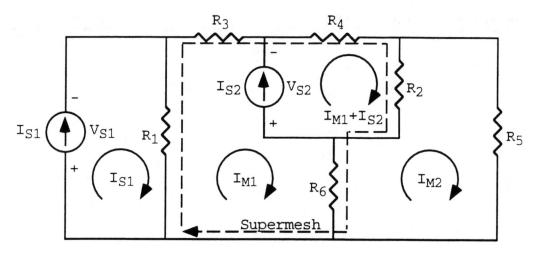

Figure 3.21 Circuit of Fig. 3.20 Showing Meshes and Supermeshes.

b) The KVL equations for meshes B and C are

$$R_1\left(I_{M1}-I_{S1}\right)+R_3 I_{M1}+R_6\left(I_{M1}-I_{M2}\right)-V_{S2}=0$$

$$R_2\left[\left(I_{M1}+I_{S2}\right)-I_{M2}\right]+R_4\left(I_{M1}+I_{S2}\right)+V_{S2}=0$$

Adding these gives

$$R_1\left(I_{M1}-I_{S1}\right)+R_3 I_{M1}+R_6\left(I_{M1}-I_{M2}\right)+R_2\left[\left(I_{M1}+I_{S2}\right)-I_{M2}\right]+R_4\left(I_{M1}+I_{S2}\right)=0$$

This equation is KVL for the sum of voltage drops around the supernode path in Fig. 3.21.

c) The remaining essential KVL mesh equation is for mesh 2 and is

$$R_2\left[I_{M2}-\left(I_{M1}+I_{S2}\right)\right]+R_5 I_{M2}+R_6\left[I_{M2}-I_{M1}\right]=0$$

d) Collecting terms gives

$$\left(R_1+R_2+R_3+R_4+R_6\right)I_{M1}-\left(R_2+R_6\right)I_{M2}=R_1 I_{S1}-\left(R_2+R_4\right)I_{S2}$$

$$-\left(R_2+R_6\right)I_{M1}+\left(R_2+R_5+R_6\right)I_{M2}=R_2 I_{S2}$$

e) Writing these two KVL equations in matrix form gives

$$\begin{bmatrix} R_1+R_2+R_3+R_4+R_6 & -\left(R_2+R_6\right) \\ -\left(R_2+R_6\right) & R_2+R_5+R_6 \end{bmatrix}\begin{bmatrix} I_{M1} \\ I_{M2} \end{bmatrix}=\begin{bmatrix} R_1 I_{S1}-\left(R_2+R_4\right)I_{S2} \\ R_2 I_{S2} \end{bmatrix}$$

$$=\begin{bmatrix} R_1 \\ 0 \end{bmatrix}I_{S1}+\begin{bmatrix} -\left(R_2+R_4\right) \\ R_2 \end{bmatrix}I_{S2}=\begin{bmatrix} R_1 & -\left(R_2+R_4\right) \\ 0 & R_2 \end{bmatrix}\begin{bmatrix} I_{S1} \\ I_{S2} \end{bmatrix}$$

```
% Example 3.8 - Supermesh Example
clear
R1=1; R2=1; R3=1; R4=1; R5=1; R6=1;
Rmat=nmAcc(0,1,R1,zeros(2,2));
Rmat=nmAcc(1,2,R2,Rmat);
Rmat=nmAcc(0,1,R3,Rmat);
Rmat=nmAcc(0,1,R4,Rmat);
Rmat=nmAcc(0,2,R5,Rmat);
Rmat=nmAcc(1,2,R6,Rmat)
As=zeros(2,2);
As(:,1)=srcAcc(0,1,R1,As(:,1));
As(:,2)=srcAcc(1,2,R2,As(:,2));
As(:,2)=srcAcc(1,0,R4,As(:,2))
Is=[1 0;0 1]
Im=Rmat\(As*Is)
```

Figure 3.22 MATLAB M-File to Solve Exercise 3.8, Part f).

f) A MATLAB M-file for this part appears in Fig. 3.22 and the numerical solution that it produces is in Fig. 3.23. When using srcAcc to build the source matrix, enter the index of the rise mesh first and the index of the drop mesh second. For example, current source I_{S2} produces a drop in mesh 1 and a rise in mesh two as it passes through resistance R_2, so the srcAcc function arguments are 1, 2, R4, and As(:,2) to accumulate the effect of R_4 onto the second column of matrix $\mathcal{A}_s$. To obtain both solutions with the same computation the M-file uses a two-column source matrix. The M-file uses literal values for the resistances to document the entry of each resistance.

```
Rmat =
     5      -2
    -2       3
As =
     1      -2
     0       1
Is =
     1       0
     0       1
Im =
   0.2727   -0.3636
   0.1818    0.0909
```

Figure 3.23 Solution for Exercise 3.8, Part f).

In Example 3.8, R_1, R_2, R_3, R_4, and R_6 are self resistances for supermesh 1, because they are in the supermesh path. Resistances R_2, R_5, and R_6 are in mesh 2, so are self resistances for mesh 2. Both R_2 and R_6 are mutual resistances between supermesh 1 and mesh 2, so they accumulate negatively in the resistance matrix in row 1, column 2 and in row 2, column 1. Also, R_1 and R_6 are in the path of two meshes so appear in the typical pattern-of-four entries, with two positive diagonal entries and two negative off-diagonal entries. The remaining resistances lie in the path of only one mesh or supermesh current, so they appear only as additive terms in the appropriate diagonal positions.

Another useful viewpoint is to think of the voltage components due to one source or mesh current at a time. For example, with I_{S1}, I_{S2}, and I_{M2} set to zero, the sum of voltages around the supermesh is

$$(R_1 + R_2 + R_3 + R_4 + R_6)I_{M1}$$

This term accounts for the entry in the first row and first column of the resistance matrix. The voltage drop about mesh 2 due to I_{M1} alone is

$$(R_2 + R_6)I_{M1}$$

which accounts for the entry in row 1, column 2 of the resistance matrix.

On the right-hand side of the matrix equation voltage rises due to the current sources appear. With I_{S1} alone not zero, there is a voltage rise in supermesh 1 equal to $R_1 I_{S1}$. This fact accounts for the nonzero entry in row 1 of the I_{S1} coefficient matrix. With I_{S2} alone having a nonzero value, resistance R_2 causes a voltage drop $R_2 I_{S2}$ in supermesh 1 and a rise in mesh 2. This fact accounts for the negative entry in row 1 and positive entry in row 2 of the I_{S2} coefficient matrix.

The point of these observations is that entry of every element in the matrix formulation of the KVL essential equations can be done by direct inspection of the circuit diagram. Writing the matrix form of the solution directly by inspection of the circuit eliminates wasting time and eliminates the possibility of error in writing and transcribing the KVL equations.

3.2.3 Mesh Analysis by Inspection

As for nodal analysis, identify the number of independent KVL equations using

$$\text{No. of Independent KVL Equations} = b - (n - 1) - n_{is} \qquad (3.32)$$

Here, b is the number of branches in the circuit and n_{is} is the number of independent and dependent current sources. For convenience, define all mesh currents using a clockwise direction of rotation. Identify the solution vector of mesh currents I_m. This vector contains as many components as the number of independent mesh currents given by Eq. 3.32. These components may be in any appropriate order. Any mesh current having a value determined by a current source on the perimeter of the circuit is known and is not a member of the solution set. Also, internal current sources establish constraints between neighboring mesh currents and define a supermesh. These constraints reduce the number of mesh currents in the solution vector. The choice of the mesh current to retain as the supermesh current from among the set of mesh currents constituting a supermesh is arbitrary but affects the form of the equations. The choice may depend on output considerations. Mark your mesh current choices on the circuit diagram and show how mesh currents not in the solution set depend on solution-set mesh currents and current sources.
Next, write the KVL mesh analysis equations

$$\mathcal{R}_{mm}I_m + \mathcal{A}_{md}X_d = \mathcal{A}_{ms}X_s \qquad (3.33)$$

where I_m is the mesh current solution vector, X_d is the vector of dependent sources, and X_s is the vector of independent sources. To have resistance matrix $\mathcal{R}_{mm}$ show the usual symmetry, write the KVL mesh equations in the same order as the order of the mesh current components in the solution vector and add voltage drops in the path direction that corresponds to the mesh current direction. In some cases these KVL equations may be supermesh equations. Matrix $\mathcal{R}_{mm}$ accounts for voltage terms in the KVL mesh equations due to circuit resistances. The $\mathcal{A}_{md}$ matrix gives the voltage terms due to any dependent sources, and the $\mathcal{A}_{ms}$ matrix accounts for terms due to independent sources. However, the controlled sources depend on the controlling variables

$$X_d = \mathcal{A}_{dc} X_c \qquad\qquad (3.34)$$

and the controlling variables depend on the mesh-current vector, dependent-source vector, and independent-source vector

$$X_c = \mathcal{A}_{cd} X_d + \mathcal{A}_{cm} I_m + \mathcal{A}_{cs} X_s \qquad\qquad (3.35)$$

As for nodal analysis, the general mesh current equations are

$$\left\{\mathcal{R}_{mm} + \mathcal{A}_{md}\left(I_d - \mathcal{A}_{dc}\mathcal{A}_{cd}\right)^{-1}\mathcal{A}_{dc}\mathcal{A}_{cm}\right\}I_m = \left\{\mathcal{A}_{ms} - \mathcal{A}_{md}\left(I_d - \mathcal{A}_{dc}\mathcal{A}_{cd}\right)^{-1}\mathcal{A}_{dc}\mathcal{A}_{cs}\right\}X_s \qquad (3.36)$$

Equations 3.33 through 3.35 also have a supermesh solution similar to Eq. 3.17.

Example 3.9 Mesh Analysis with VCVS and VCCS Example

For the circuit shown in Fig. 3.24, which has a VCVS and a VCCS controlled source

a) Prepare the circuit for mesh analysis by inspection. Mark the meshes and supermeshes with appropriate mesh currents and indicate any supermesh paths.

b) Write the essential KVL mesh or supermesh equations in terms of the mesh currents, dependent sources, and independent source. Rearrange these, collect coefficients and move the independent source term(s) to the right-hand side. Write these in the matrix form of Eq. 3.33.

c) Write the dependency relations for I_{D1} and V_{D2} in the form of Eq. 3.34.

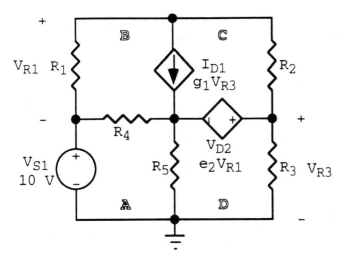

Figure 3.24 Circuit for Example 3.9 ($R_1 = 1\ \Omega$, $R_2 = 2\ \Omega$, $R_3 = 3\ \Omega$, $R_4 = 4\ \Omega$, $R_5 = 5\ \Omega$, $g_1 = 10$ A/V, $e_2 = 5$ V/V).

d) Write the controlling relations for V_{R1} and V_{R3} in terms of the dependent sources, mesh currents, and independent sources. Express these in the form of matrix Eq. 3.35. Examine the matrix form and convince yourself that you can write this equation by inspection of the circuit.

e) Use MATLAB to calculate the node-voltage vector using both the matrix analysis result, Eq. 3.36 and the super-matrix method, Eq. 3.17.

Solution

a) The circuit has b = 8, n = 4, and n_{is} = 1, so

$$\text{No. of Independent KVL Equations} = b - n - n_{is} = 8 - 4 - 1 = 3$$

Figure 3.25 shows selections for the three mesh currents. Mesh current I_{M1} appears in the A window, I_{M2} in the B and C window, and I_{M3} in the D window. The mesh choices have dependent source I_{D1} circulating on the B window.

b) The essential mesh and supermesh equations are

$$R_4\left[I_{M1} - \left(I_{M2} + I_{D1}\right)\right] + R_5\left(I_{M1} - I_{M3}\right) - V_{S1}$$

$$R_1\left(I_{M2} + I_{D1}\right) + R_2 I_{M2} + V_{D2} + R_4\left[\left(I_{M2} + I_{D1}\right) - I_{M1}\right] = 0$$

$$R_3 I_{M3} + R_5\left(I_{M3} - I_{M1}\right) - V_{D2} = 0$$

Collecting coefficients and moving independent source terms to the right-hand side

$$\left(R_4 + R_5\right)I_{M1} - R_4 I_{M2} - R_5 I_{M3} - R_4 I_{D1} = V_{S1}$$

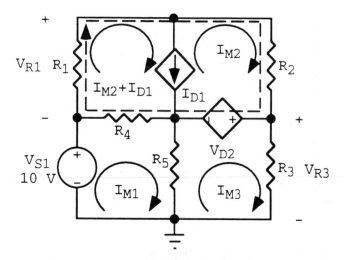

Fig. 3.25 Circuit for Example 3.9 with Mesh and Supermesh Definitions.

$$-R_4 I_{M1} + \left(R_1 + R_2 + R_4\right) I_{M2} + \left(R_1 + R_4\right) I_{D1} + V_{D2} = 0$$

$$-R_5 I_{M1} + \left(R_3 + R_5\right) I_{M3} - V_{D2} = 0$$

Writing In matrix form or directly from the circuit

$$\begin{bmatrix} R_4 + R_5 & -R_4 & -R_5 \\ -R_4 & R_1 + R_2 + R_4 & 0 \\ -R_5 & 0 & R_3 + R_5 \end{bmatrix} \begin{bmatrix} I_{M1} \\ I_{M2} \\ I_{M3} \end{bmatrix} + \begin{bmatrix} -R_4 & 0 \\ R_1 + R_4 & 1 \\ 0 & -1 \end{bmatrix} \begin{bmatrix} I_{D1} \\ V_{D2} \end{bmatrix} = \begin{bmatrix} 1 \\ 0 \\ 0 \end{bmatrix} V_{S1}$$

The resistance coefficient matrix $\mathscr{R}_{mm}$ shows the self- and mutual-resistance terms, which appear in familiar pattern-of-four arrays. The coefficients of the first column of the $\mathscr{A}_{md}$ matrix that multiplies the dependent-source matrix are the ratios of net voltage in each of the three mesh equations to I_{D1}. For example, $R_1 + R_4$ is the ratio of the voltage drops about the second, supermesh equation to current I_{D1}. The coefficients of the second column of the $\mathscr{A}_{md}$ matrix are the ratios of net voltage in each of the three mesh equations to V_{D2}. The coefficients of the $\mathscr{A}_{ms}$ matrix that multiplies V_{S1} give the ratios of the net voltage rise about each mesh to voltage source V_{S1}.

c) The matrix equation that relates the controlled sources and controlling variables is

$$\begin{bmatrix} I_{D1} \\ V_{D2} \end{bmatrix} = \begin{bmatrix} g_1 & 0 \\ 0 & e_2 \end{bmatrix} \begin{bmatrix} V_{R3} \\ V_{R1} \end{bmatrix}$$

d) The controlling variables are resistance voltages and Ohm's law gives these in terms of mesh currents

$$V_{R1} = R_1 \left(I_{M2} + I_{D1}\right)$$

$$V_{R3} = R_3 I_{M3}$$

In matrix form or directly from the circuit

$$\begin{bmatrix} V_{R3} \\ V_{R1} \end{bmatrix} = \begin{bmatrix} 0 & 0 \\ -R_1 & 0 \end{bmatrix} \begin{bmatrix} I_{D1} \\ V_{D2} \end{bmatrix} + \begin{bmatrix} 0 & 0 & R_3 \\ 0 & -R_1 & 0 \end{bmatrix} \begin{bmatrix} I_{M1} \\ I_{M2} \\ I_{M3} \end{bmatrix} + \begin{bmatrix} 0 \\ 0 \end{bmatrix} V_{S1}$$

which displays coefficient matrices $\mathscr{A}_{cd}$, $\mathscr{A}_{cm}$, and $\mathscr{A}_{cs}$.

e) The MATLAB M-file in Fig. 3.26 calculates the mesh-current vector using matrix Eq. 3.36 and the supermesh method Eq. 3.17, and the solution that MATLAB finds is in Fig. 3.27. Both solutions show that $I_{M1} = 2.0179$ A, $I_{M2} = 4.0358$ A, and $I_{M3} = -0.0639$ A. The supermatrix method also gives $I_{D1} = -1.9157$ A, $V_{D2} = -10.6003$ V, $V_{R3} = -0.1916$ V, and $V_{R1} = -2.1201$ V

```
% Example 3.9 - Mesh Analysis with VCVS and VCCS Example
clear, format compact
R1=1; R2=2; R3=3; R4=4; R5=5;
g1=10;        e2=5;
% Define resistance matrix
Rmm=nmAcc(0,2,R1,zeros(3));
Rmm=nmAcc(0,2,R2,Rmm);
Rmm=nmAcc(0,3,R3,Rmm);
Rmm=nmAcc(1,2,R4,Rmm);
Rmm=nmAcc(1,3,R5,Rmm);
% Define Amd
Amd=zeros(3,2);
Amd(:,1)=srcAcc(0,2,1,Amd(:,1));
Amd(:,1)=srcAcc(1,2,4,Amd(:,1));
Amd(:,2)=srcAcc(3,2,1,Amd(:,2));
% Define remaining matrices
Ams=srcAcc(0,1,1,zeros(3,1));
Adc=[g1 0;0 e2];
Acd=[0 0;-1 0];
Acm=[0 0 3;0 -1 0];
Acs=zeros(2,1);
Vs1=10; % Define Source Vector
disp('Mesh-current vector Im using genAnal:')
Im=genAnal(Vs1,Rmm,Amd,Ams,Adc,Acd,Acm,Acs);
disp(Im)
disp('Super matrix Rsuper:')
[rRmm,cRmm]=size(Rmm)
[rAdc,cAdc]=size(Adc)
[rAcd,cAcd]=size(Acd);
Zmc=zeros(rRmm,cAdc);, Zdm=zeros(rAdc,cRmm);
Ud=eye(rAdc);, Uc=eye(rAcd);
Rsuper=[Rmm Amd Zmc;Zdm Ud -Adc;-Acm -Acd Uc];
disp(Rsuper)
XSuper=[Ams;zeros(rAdc,1);Acs];
Ysuper=Rsuper\XSuper*Vs1;
disp('Im, Xd, and Xc')
disp(Ysuper)
```

Figure 3.26 MATLAB M-File for Example 3.9.

```
Mesh-current vector Im using genAnal:
    2.0179
    4.0358
   -0.0639
Super matrix Rsuper:
      9     -4     -5     -4      0      0      0
     -4      7      0      5      1      0      0
     -5      0      8      0     -1      0      0
      0      0      0      1      0    -10      0
      0      0      0      0      1      0     -5
      0      0     -3      0      0      1      0
      0      1      0      1      0      0      1
Im, Xd, and Xc
    2.0179
    4.0358
   -0.0639
   -1.9157
  -10.6003
   -0.1916
   -2.1201
```

Figure 3.27 Solution for Example 3.9.

3.2.4 Mutual Inductance

The previous discussion assumes that the circuits operate at DC or are slowly varying. Of course, the analysis procedures above can be done when the circuit is a phasor circuit. Then some elements have impedance or admittance scalar relationships between their phasor voltages and currents. Of particular interest now is an effective procedure to deal with inductance elements that have mutual coupling.

Entry of mutual inductance terms directly into mesh equations by inspection of the circuit often is possible, however the matrix approach shown here simplifies the work. This approach removes inductance voltages from the normal impedance matrix, placing these terms instead into a separate matrix. Then, with impedance matrix Z_{mm} replacing resistance matrix $\mathcal{R}_{mm}$, Eq. 3.32 becomes

$$Z_{mm}I_m + \mathcal{A}_{md}X_d + \mathcal{A}_{m\ell}V_\ell = \mathcal{A}_{ms}X_s \qquad (3.37)$$

In Eq. 3.37 the third term $\mathcal{A}_{m\ell}V_\ell$ accounts for inductance voltage components. The vector of controlled sources depend on the vector of controlled variables as in Eq. 3.34, repeated here

$$X_d = \mathcal{A}_{dc}X_c \qquad (3.38)$$

and the controlling variables depend on mesh currents and all sources as in Eq. 3.35, also repeated here.

$$X_c = \mathcal{A}_{cd}X_d + \mathcal{A}_{cm}I_m + \mathcal{A}_{cs}X_s \qquad (3.39)$$

The inductance voltages depend on inductance currents so

$$\mathcal{V}_\ell = s\mathcal{L}I_\ell \tag{3.40}$$

If a single dot set describes the mutual inductance polarity relationships and all inductance voltages and currents have polarities consistent with the dot convention (positive voltage at and current into the dotted end), then inductance matrix $\mathcal{L}$ has all positive coefficient values. Finally, inductance current vector I_ℓ depends linearly on the dependent-source, mesh, and independent-source vectors, so

$$I_\ell = \mathcal{A}_{\ell d}X_d + \mathcal{A}_{\ell m}I_m + \mathcal{A}_{\ell s}X_s \tag{3.41}$$

Through matrix partitioning, the solution to Eqs. 3.37 through 3.41 has the same form as Eq. 3.36. First, write Eq. 3.37 as

$$Z_{mm}I_m + \begin{bmatrix} \mathcal{A}_{md} & \vdots & \mathcal{A}_{m\ell} \end{bmatrix}\begin{bmatrix} X_d \\ \hline \mathcal{V}_\ell \end{bmatrix} = Z_{mm}I_m + \mathcal{A}'_{md}X'_d = \mathcal{A}_{ms}X_s \tag{3.42}$$

where X'_d is a composite vector constructed from X_d and $\mathcal{V}_\ell$. Next, combine Eq. 3.37 and 3.41 in the form

$$X'_d = \begin{bmatrix} X_d \\ \hline \mathcal{V}_\ell \end{bmatrix} = \mathcal{A}'_{dc}X'_c = \begin{bmatrix} \mathcal{A}_{dc} & \vdots & 0 \\ \hline 0 & \vdots & s\mathcal{L} \end{bmatrix}\begin{bmatrix} X_c \\ \hline I_\ell \end{bmatrix} \tag{3.43}$$

Finally, combine Eq. 3.39 and 3.40 as

$$\begin{aligned}
X'_c = \begin{bmatrix} X_c \\ \hline I_\ell \end{bmatrix} &= \mathcal{A}'_{cd}X'_d + \mathcal{A}'_{cm}I_m + \mathcal{A}'_{cs}X_s \\
&= \begin{bmatrix} \mathcal{A}_{cd} & \vdots & 0 \\ \hline \mathcal{A}_{\ell d} & \vdots & 0 \end{bmatrix}\begin{bmatrix} X_d \\ \hline \mathcal{V}_\ell \end{bmatrix} + \begin{bmatrix} \mathcal{A}_{cm} \\ \hline \mathcal{A}_{\ell m} \end{bmatrix}I_m + \begin{bmatrix} \mathcal{A}_{cs} \\ \hline \mathcal{A}_{\ell s} \end{bmatrix}X_s
\end{aligned} \tag{3.44}$$

Solving Eqs. 3.42 through 3.44 gives

$$\left\{ Z_{mm} + \mathcal{A}'_{md}\left(I - \mathcal{A}'_{dc}\mathcal{A}'_{cd}\right)^{-1}\mathcal{A}'_{dc}\mathcal{A}'_{cm} \right\}I_m = \left\{ \mathcal{A}_{ms} - \mathcal{A}'_{md}\left(I - \mathcal{A}'_{dc}\mathcal{A}'_{cd}\right)^{-1}\mathcal{A}'_{dc}\mathcal{A}'_{cs} \right\}X_s \tag{3.45}$$

Example 3.10, using the circuit in Fig. 3.28 and having mutual inductance M between inductances L_1 and L_2, shows how to write the KVL mesh and ancillary equations by inspection.

Example 3.10 Mutual Inductance Example

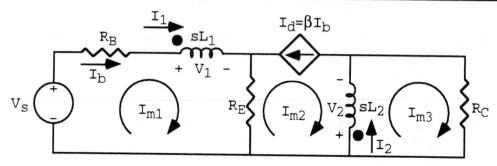

Figure 3.28 Mutual Inductance Example (Inductances L_1 and L_2 coupled with mutual inductance M).

Write the three basic matrix equations 3.42, 3.43, and 3.44 for the circuit of Fig. 3.28

Solution

Mesh Eq. 3.42 is

$$\begin{bmatrix} R_B + R_E & 0 \\ 0 & R_C \end{bmatrix}\begin{bmatrix} I_{m1} \\ I_{m3} \end{bmatrix} + \begin{bmatrix} R_E & 1 & 0 \\ 0 & 0 & 1 \end{bmatrix}\begin{bmatrix} I_d \\ \overline{V_{l1}} \\ V_{l2} \end{bmatrix} = \begin{bmatrix} 1 \\ 0 \end{bmatrix}[V_s]$$

Next, write Eq. 3.43

$$\begin{bmatrix} I_d \\ \overline{V_{l1}} \\ V_{l2} \end{bmatrix} = \begin{bmatrix} \beta & 0 & 0 \\ 0 & sL_1 & sM \\ 0 & sM & sL_2 \end{bmatrix}\begin{bmatrix} I_b \\ \overline{I_{l1}} \\ I_{l2} \end{bmatrix}$$

Writing Eq. 3.44

$$\begin{bmatrix} I_b \\ \overline{I_{l1}} \\ I_{l2} \end{bmatrix} = \begin{bmatrix} 0 & 0 & 0 \\ 0 & 0 & 0 \\ 1 & 0 & 0 \end{bmatrix}\begin{bmatrix} I_d \\ \overline{V_{l1}} \\ V_{l2} \end{bmatrix} + \begin{bmatrix} 1 & 0 \\ 1 & 0 \\ 0 & 1 \end{bmatrix}\begin{bmatrix} I_{m1} \\ I_{m3} \end{bmatrix} + \begin{bmatrix} 0 \\ 0 \\ 0 \end{bmatrix}[V_s]$$

The solution of Eqs 3.42, 3.43, and 3.44 now follows using Eq. 3.45.

3.3 Modified Nodal Analysis

Modified nodal analysis (MNA) is a method for formulating circuit equations that is simpler to write than nodal analysis using supernodes. This method is easy to implement with a computer program, so is the analysis method that many computer circuit analysis programs use. It also is an effective starting point for MATLAB circuit analysis.

3.3.1 Introduction

The MNA method resolves all circuit elements into three groups. The first group contains elements that have a natural nodal analysis representation and whose currents do not need to appear in the solution. The second group consists of elements that do not have a nodal analysis representation or whose currents are to appear in the solution. The last group includes all independent current sources. With this grouping, each of the n node KCL equations has the form

$$\sum_{j=1}^{n_1} i_{1_j} + \sum_{j=1}^{n_2} i_{2_j} = \sum_{j=1}^{n_3} i_{3_j} \tag{3.46}$$

The currents in group 1 enter each KCL equation in terms of the admittance of each group 1 element times the node-voltage difference for the nodes that the element connects. Each element in the second group introduces it's current as a component in the solution vector. At the same time, each element in the second group contributes a constraint equation. Each additional constraint equation, known as a constitutive equation (CE), increases the number of equations in the solution set by one. The complete set of equations has the form

$$
\begin{array}{c}
\text{KCL Equations} \left\{ \\ \\ \\ \text{Constitutive} \\ \text{Equations} \left\{ \\ \\ \\
\end{array}
\begin{bmatrix}
G_{11} & \cdots & G_{1n} & A_{11} & \cdots & A_{1m} \\
\vdots & \ddots & \vdots & \vdots & \ddots & \vdots \\
G_{n1} & \cdots & G_{nn} & A_{n1} & \cdots & A_{nm} \\
\hdashline
X_{11} & \cdots & X_{1n} & B_{11} & \cdots & B_{1m} \\
\vdots & \ddots & \vdots & \vdots & \ddots & \vdots \\
X_{m1} & \cdots & X_{mn} & B_{m1} & \cdots & B_{mm}
\end{bmatrix}
*
\begin{bmatrix}
V_{n1} \\ \vdots \\ V_{nn} \\ \hdashline I_{x1} \\ \vdots \\ I_{xm}
\end{bmatrix}
=
\begin{bmatrix}
J_1 \\ \vdots \\ J_n \\ \hdashline W_1 \\ \vdots \\ W_m
\end{bmatrix}
$$

$$(3.47)$$

The following examples will help you to understand how to construct Eq. (3.47).

Example 3.11 MNA Equations with Independent Voltage Source

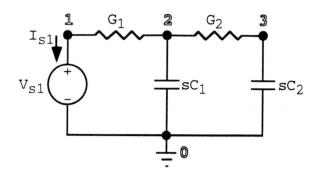

Figure 3.29 Circuit for Modified Nodal Analysis with Voltage Source ($G_1 = G_2 =$ 1 mS, $C_1 = C_2 = 1\,\mu F$, $s = j$ krad/s, $V_{s1} = 1\angle 0$ V).

Write the MNA matrix equation for the circuit in Fig. 3.29.

Solution

The voltage source adds current I_{s1} as a component of the solution vector and also provides the CE

$$V_{n1} = V_{s1}$$

Writing Eq. 3.47 for this example

$$
T(s)X =
\begin{bmatrix}
G_1 & -G_1 & 0 & 1 \\
-G_1 & sC_1 + G_1 + G_2 & -G_2 & 0 \\
0 & -G_2 & sC_2 + G_2 & 0 \\
\hdashline
1 & 0 & 0 & 0
\end{bmatrix}
\begin{bmatrix}
V_{n1} \\ V_{n2} \\ V_{n3} \\ \hdashline I_{s1}
\end{bmatrix}
=
\begin{bmatrix}
0 \\ 0 \\ 0 \\ \hdashline V_{s1}
\end{bmatrix}
= W
$$

The elements in the fourth row of matrix $T(s)$ account for the CE and the first three elements in the fourth column of $T(s)$ incorporate current I_s into the three KCL equations.

The `help` documentation for a new MATLAB function `mnaVSrc` is in Fig. 3.30. This function simplifies entry of independent voltage source terms into MNA equations. Using the M-file in Fig. 3.31 to solve Example 3.11 gives the results in Fig. 3.32. Notice that the M-file expands the circuit matrix T into a 4×4 matrix

```
amat=mnaVSrc(m,i,j,bmat)
 Accumulation of Vsrc onto bmat for MNA analysis.
   m is the index into the additional CE row.
   i and j are the + and - nodes of the voltage source.
```

Figure 3.30 mnaVSrc Help.

```
% Example 3.11 - MNA Equations with V Source
clear
G1=1;, G2=1;
C1=1;, C2=1;
s=j;
T=nmAcc(1,2,G1,zeros(3));,        T=nmAcc(2,3,G2,T);
T=nmAcc(2,0,s*C1,T);,             T=nmAcc(3,0,s*C2,T);
T=mnaVSrc(4,1,0,T)
W=[0; 0; 0; 1];
y=T\W
```

Figure 3.31 M-File to Solve Example 3.11.

```
T =
    1.0000            -1.0000                    0            1.0000
   -1.0000             2.0000 + 1.0000i  -1.0000                   0
        0             -1.0000             1.0000 + 1.0000i         0
    1.0000                  0                    0                 0
y =
    1.0000
    0.3333 - 0.3333i
    0.0000 - 0.3333i
   -0.6667 - 0.3333i
```

Figure 3.32 MATLAB Solution of Example 3.11.

by referring to the 4-th row and column in the first argument of mnaVSrc, after first defining T as a 3×3 zero-valued matrix.

Example 3.11 shows that solving a circuit problem using MNA may result in more equations than nodal analysis, but that additional solution variables appear in the solution vector. From the viewpoint of MNA, independent or dependent voltage sources do not reduce the number of KCL node equations. The following section shows how to incorporate controlled sources and various other two-port circuit elements into the MNA method.

3.3.2 Patterns for Type-Two Elements in MNA

Each type-two element requires a different treatment in modified nodal analysis. The following sections develop the procedure for common type-two elements from their constitutional equations and the current terms that each introduce into the KCL equations.

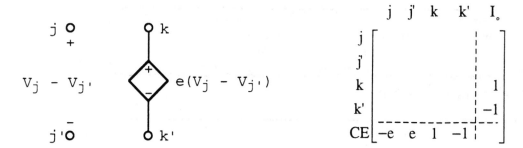

a) E Source.   b) MNA Matrix Representation.

Figure 3.33 MNA Representation of E Source.

```
amat=mnaESrc (m,j,jp,k,kp,Eval,bmat)
  Accumulation of Esrc having value "Eval" onto bmat.
  m is the index of the added CE row.
  j and jp are the + and - controlling nodes.
  k and kp are the + and - nodes of the E source.
```

Figure 3.34 mnaESrc Help.

E Source (VCVS)

The voltage-controlled voltage source (VCVS) shown in Fig. 3.33a has the additional CE

$$V_k - V_{k'} - e(V_j - V_{j'}) = 0 \qquad (3.48)$$

Figure 3.33b shows the entries in the additional CE row and the additional current column. The new MATLAB function mnaESrc, having the help documentation in Fig. 3.34, simplifies construction of the MNA equations when a circuit contains a VCVS. Function mnaESrc accumulates Eval onto the m row in the j and j′ column to avoid eliminating the 1 or -1 if j or j′ happen to be the same as k or k′.

F Source (CCCS)

The current-controlled current source (CCCS) shown in Fig. 3.35a has two additional CEs

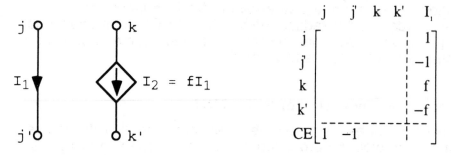

a) F Source.   b) MNA Matrix Representation.

Figure 3.35 MNA Representation of F Source.

```
amat=mnaFSrc (m,j,jp,k,kp,Fval,bmat)
  Accumulation of Fsrc value "Fval" onto bmat.
  m is the index of the added CE row.
  j and jp are the from and to controlling nodes.
  k and kp are the from and to nodes of the F source.
```

Figure 3.36 mnaFSrc Help.

$$V_j - V_{j'} = 0 \qquad\qquad (3.49)$$

$$I_2 - fI_1 = 0 \qquad\qquad (3.50)$$

However, Fig. 3.35b shows that current I_2 does not have to appear in the solution set. The I_1 column includes f and -f to account for the I_2 current in the KCL portion of the MNA equations. The one additional CE row gives the CE of the short-circuit input branch. The new MATLAB function, having the `help` information in Fig. 3.36, simplifies construction of the MNA equation for a circuit having a CCCS. Instead of setting Fval, function `mnaFSrc` accumulates `Fval` into the matrix, in case k or k' are the same as either j or j'. Example 3.12 uses mnaESrc to solve a circuit problem and avoids a trap by not using `mnaFSrc`. To use both `mnaESrc` and `mnaFSrc` with this example, insert a null-valued voltage source in series with E1 to measure F2's current.

Example 3.12 Circuit with F and E Sources

For the circuit shown in Fig. 3.37, compute the node voltages V_{n1}, V_{n2}, ..., V_{n5} and currents I_{vs} and I_1.

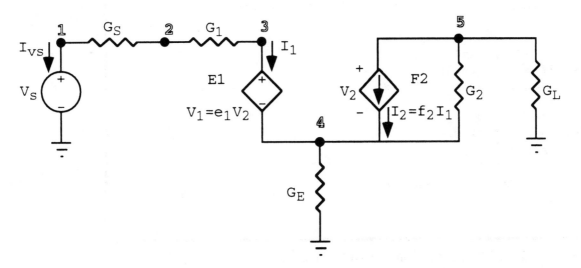

Figure 3.37 Circuit for Example 3.12 ($G_1 = G_S = G_E = 1$ mS, $G_L = 0.2$ mS, $G_2 = 10\ \mu$S, $e_1 = 10^{-3}$, $f_2 = 50$).

Solution

The MNA equation for this circuit is

$$
\begin{bmatrix}
G_S & -G_S & 0 & 0 & 0 & | & 1 & 0 \\
-G_S & G_S+G_1 & -G_1 & 0 & 0 & | & 0 & 0 \\
0 & -G_1 & G_1 & 0 & 0 & | & 0 & 1 \\
0 & 0 & 0 & G_E+G_2 & -G_2 & | & 0 & -1-f_2 \\
0 & 0 & 0 & -G_2 & G_2+G_L & | & 0 & f_2 \\
\hline
1 & 0 & 0 & 0 & 0 & | & 0 & 0 \\
0 & 0 & 1 & -1+e_1 & -e_1 & | & 0 & 0
\end{bmatrix}
*
\begin{bmatrix}
V_{n1} \\
V_{n2} \\
V_{n3} \\
V_{n4} \\
V_{n5} \\
\hline
I_{vs} \\
I_1
\end{bmatrix}
=
\begin{bmatrix}
0 \\
0 \\
0 \\
0 \\
0 \\
\hline
V_s \\
0
\end{bmatrix}
$$

The M-file in Fig. 3.38 solves this matrix equation, but does not use the `mnaFSrc` function to enter the CCCS values into the circuit matrix. If we use `mnaFSrc` here, -1 replaces the entry $-1+e_1$ in row 7, column 4. The additional f_2 entries accumulate easily into column 7, rows 4 and 5. The conductance matrix for this circuit and the solution vector that MATLAB calculates appear in Fig. 3.39. Another solution that uses `mnaFSrc` uses the source V_S to measure the current that controls F2. In this solution replace

```
G([4;5],7)=G([4;5],7)+f2*[-1;1]
```

in the M-file of Fig. 3.38 with

```
G=mnaFSrc(6,1,0,5,4,-f2,G)
```

The sign of the f2 term here is negative, because $I_{VS} = -I_1$.

```
% Example 3.12 - Circuit with F and E Sources
clear
Vs=1;
GS=1;  G1=1;  GE=1;  G2=1/100;    GL=1/5;
e1=1e-3;       f2=50;
G=nmAcc(1,2,GS,zeros(5));
G=nmAcc(2,3,G1,G);
G=nmAcc(4,0,GE,G);
G=nmAcc(5,4,G2,G);
G=nmAcc(5,0,GL,G);
G=mnaVSrc(6,1,0,G);
G=mnaESrc(7,5,4,3,4,e1,G);
G([4;5],7)=G([4;5],7)+f2*[-1;1]
W=[0 0 0 0 0 Vs 0]';
X=G\W
```

Figure 3.38 M-File to Solve Example 3.12.

```
G =
     1.0000    -1.0000         0         0         0    1.0000         0
    -1.0000     2.0000    -1.0000         0         0         0         0
          0    -1.0000     1.0000         0         0         0    1.0000
          0          0         0    1.0100   -0.0100         0   -51.0000
          0          0         0   -0.0100    0.2100         0    50.0000
     1.0000          0         0         0         0         0         0
          0          0    1.0000   -0.9990   -0.0010         0         0

X =
     1.0000
     0.9800
     0.9599
     0.9656
    -4.7277
    -0.0200
     0.0200
```

Figure 3.39 Conductance Matrix and Solution Vector for Ex. 3.12.

H Source (CCVS)

The current-controlled voltage source (CCVS) shown in Fig. 3.40a has two additional CEs

$$V_j - V_{j'} = 0 \tag{3.51}$$

$$V_k - V_{k'} - hI_1 = 0 \tag{3.52}$$

Figure 3.40b shows the entries in the additional CE rows and the two additional current columns. The new MATLAB function in Fig. 3.41 simplifies construction of a circuit system equation for a circuit having a CCVS.

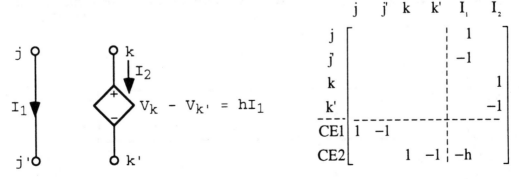

a) H Source.   b) MNA Matrix Representation.

Figure 3.40 MNA Representation of H Source.

```
amat=mnaHSrc(m,j,jp,k,kp,Hval,bmat)
   Accumulation of Hsrc value "Hval" onto bmat.
   m is the index of the added CE row.
   j and jp are the from and to controlling nodes.
   k and kp are the + and - nodes of the H source.
```

Figure 3.41 mnaHSrc Help.

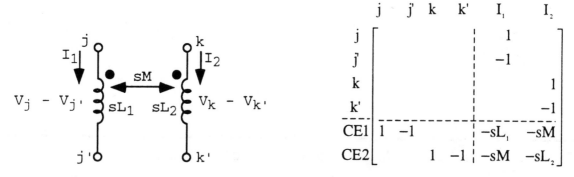

a) Mutually-Coupled Inductances.   b) MNA Matrix Representation.

Figure 3.42 MNA Representation of Mutually-Coupled Inductances.

Mutually-Coupled Inductances

The mutually-coupled inductances in Fig. 3.42a provide the additional CEs

$$V_j - V_{j'} - sL_1 I_1 - sMI_2 = 0 \tag{3.53}$$

$$V_k - V_{k'} - sMI_1 - sL_2 I_2 = 0 \tag{3.54}$$

Figure 3.42b shows the entries in the additional CE rows and the two additional current columns. A new MATLAB function `mnaLM`, having the `help` documentation in Fig. 3.43, simplifies construction of the MNA equation for a circuit having mutually-coupled inductances. Use this function and `mnaVSrc` to enter all of the values. Example 3.13 shows how to set up the MNA equations in MATLAB for a circuit having mutual inductance coupling of two inductances.

```
amat=mnaLM(m,L1,L2,M,bmat)
   Accumulation of L1, L2, and M onto bmat
   m is the index of the first of two additional CE rows.
```

Figure 3.43 mnaLM Help.

Example 3.13 Circuit with Mutual Inductance

a) For the circuit shown in Fig. 3.44, determine the value of mutual inductance M that gives series resonance at frequency ω = 1 Mrad/s

b) With the mutual inductance value from part a), plot the Bode dB and phase of current I_1 vs. ω, for values of ω ranging from 0.1 Mrad/s to 10 Mrad/s.

Solution

a) The MNA equations for the circuit in Fig. 3.45 are

$$
\begin{bmatrix}
G_1 & -G_1 & 0 & 1 & 0 & 0 \\
-G_1 & G_1 & 0 & 0 & 1 & 0 \\
0 & 0 & sC_1 & 0 & 0 & 1 \\
\hline
1 & 0 & 0 & 0 & 0 & 0 \\
0 & 1 & 0 & 0 & -sL_1 & -sM \\
0 & 0 & 1 & 0 & -sM & -sL_2
\end{bmatrix}
*
\begin{bmatrix}
V_{n1} \\
V_{n2} \\
V_{n3} \\
\hline
I_{in} \\
I_1 \\
I_2
\end{bmatrix}
=
\begin{bmatrix}
0 \\
0 \\
0 \\
\hline
V_{in} \\
0 \\
0
\end{bmatrix}
$$

The value of M cannot b e greater than $\sqrt{L_1 L_2}$, which equals 2 mH, so the M-file solution in Fig. 3.45 varies M from 0 to 2 mH in 0.1-mH steps. Values in the M-file are in the units V, mA, μs, Mrad/s, kΩ, mH, and nF. The for loop computes 21 values of current I_1 for each value of mutual inductance. After completing the computations, the plot command gives the plot of I_1 vs. M, shown in Fig. 3.46. Using MATLAB's `ginput` function

```
»[Mval,Phase]=ginput(1)
»Mval
Mval =1.7307
```

shows that the phase shift of current I_1 becomes zero at ω = 1 Mrad/s when M $\approx$ 1.7307 mH. This value is within 0.1% of the theoretical value for zero phase shift, which is M = $\sqrt{3}$ mH .

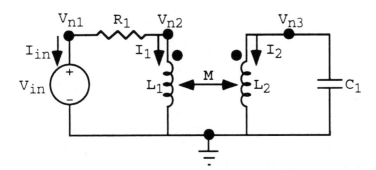

Figure 3.44 Circuit with Mutual Inductance (R_1 = $1/G_1$ = 1 kΩ, L_1 = 1 mH, L_2 = 4 mH, C_1 = 1 nF, V_{in} = $1\angle 0°$ V).

```
% Example 3.13 - Part a - Circuit with Mutual Inductance
% Using V, mA, us, Mrad/s, kOhm, mH, and nF.
clear
Vs=1;
G1=1; L1=1; L2=4; C1=1;
W=[0 0 0 Vs 0 0]';
G=nmAcc(1,2,G1,zeros(6));
G=mnaVSrc(4,1,0,G);
G=mnaVSrc(5,2,0,G);
G=mnaVSrc(6,3,0,G);
C=nmAcc(3,0,C1,zeros(6));
np=21;
w=1;
s=j*w;
M=linspace(0,2,np);
% Generate array Iin as M varies
for i=1:np
     C=mnaLM(5,L1,L2,M(i),C);
     T=s*C+G;
     X=T\W;
     I1(i)=X(5);
end
plot(M,180*angle(I1)/pi)
grid
xlabel('Mutual Inductance, mH')
ylabel('Phase of I1, Degrees')
```

Figure 3.45 M-File Solution to Example 3.13.

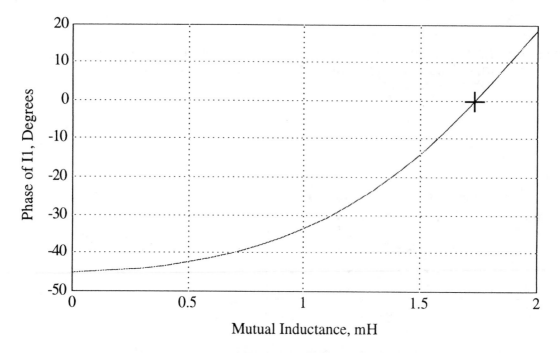

Figure 3.46 Phase Shift vs. Mutual Inductance.

```
% Example 3.13 - Part b - Circuit with Mutual Inductance
% Using V, mA, us, Mrad/s, kOhm, mH, and nF.
clear
Vs=1; G1=1; L1=1; L2=4; C1=1; M=1.7307;
W=[0 0 0 Vs 0 0]';
G=nmAcc(1,2,G1,zeros(6));        G=mnaVSrc(4,1,0,G);
G=mnaVSrc(5,2,0,G);              G=mnaVSrc(6,3,0,G);
C=nmAcc(3,0,C1,zeros(6));        C=mnaLM(5,L1,L2,M,C);
np=101;
w=logspace(-1,1,np);             s=j*w;
% Generate array I1 as w varies
for i=1:np
  T=s(i)*C+G;
  X=T\W;
  I1(i)=X(5);
end
semilogx(w,20*log10(I1))
grid
xlabel('Frequency, Mrad/s')
ylabel('Magnitude of I1, dB')
```

a) M-File for Magnitude Plot.

```
semilogx(w,180*angle(I1)/pi)
grid
xlabel('Frequency, Mrad/s')
ylabel('Phase of I1, Degrees')
```

b) Modification of M-File for Phase Plot.

Figure 3.47 M-Files for Bode Plot of I_1 for Mutual Inductance Circuit.

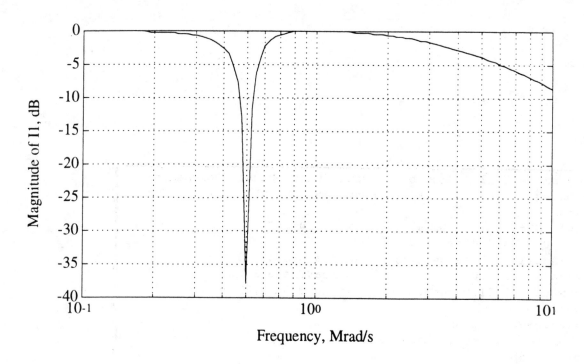

a) Magnitude.

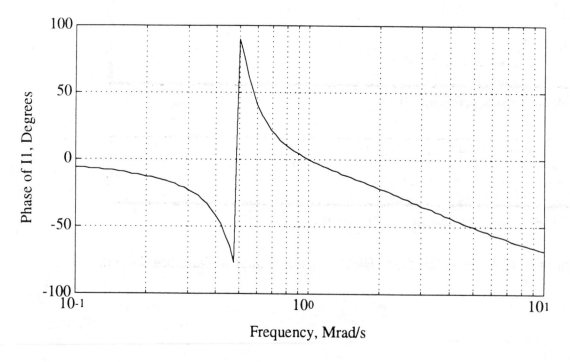

b) Phase.

Figure 3.48 Bode Magnitude and Phase Plots for Mutual Inductance Circuit.

b) To plot the Bode magnitude and phase of current I_1 vs. $\log(\omega)$, use the M-file shown in Figs. 3.47 a) and b). These M-files set up capacitance matrix C and conductance matrix G, using functions `nmAcc`, `mnaVSrc`, and `mnaLM`. The magnitude plot, which appears in Fig. 3.48a, shows that the magnitude of I_1 equals one at both zero frequency and 1 Mrad/s, since the impedance Z_1 looking into terminals 1 and 0 towards inductance L_1 equals zero at both of these frequencies. For the same reason, the phase plot in Fig. 3.48b shows that current I_1 is in phase with voltage V_{in} at both of these frequencies. The magnitude and phase plots also show that Z_1 becomes infinite and changes sign at $\omega = 0.5$ Mrad/s. Can you explain this behavior?

Ideal Transformer

The ideal transformer shown in Fig. 3.49a has two additional CEs. These are

$$I_1 + nI_2 = 0 \tag{3.55}$$

$$V_k - V_{k'} - nV_j + nV_{j'} = 0 \tag{3.56}$$

Figure 3.49b shows the entries in the additional CE rows and the two additional current columns. A new MATLAB function `mnaIdX`, which has the `help` documentation shown in Fig. 3.50, simplifies construction of a circuit system equation for a circuit having an ideal transformer. Notice that the 1 and -1

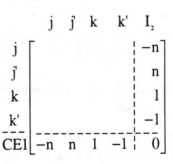

a) Ideal Transformer.

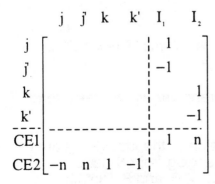

b) MNA Representation.   c) One-Current Matrix Representation.

Figure 3.49 MNA Representation of Ideal Transformer.

```
amat=mnaIdX(m,j,jp,k,kp,nVal,bmat)
   Accumulation of ideal transformer with turns ratio nVal
   onto bmat. This function adds two rows to your matrix.
     m is the index of the first additional CE row.
     j (dot) and jp are the from and to input branch nodes.
     k (dot) and kp are the from and to output branch nodes.
```

Figure 3.50 mnaIdX Help.

```
amat=mnaIdX1(m,j,jp,k,kp,nVal,bmat)
   Accumulation of ideal transformer with turns ratio nVal
   onto bmat. This function adds only one additional row
   to your matrix.
     m is the index of the one additional row.
     j (dot) and jp are the from and to input branch nodes.
     k (dot) and kp are the from and to output branch nodes.
```

Figure 3.51 mnaIdX1 Help.

values in row m+1, columns k and k+1 accumulate to avoid removing the n and
-n values if either j or j' are the same as either k or k'. The new MATLAB
function mnaIdX1, with help documentation in Fig. 3.51, allows for entry of
values for an ideal transformer into MNA equations, where only the output-port
current appears in the solution vector. Fig. 3.49c shows the matrix entries for
this formulation. Again, care is taken to avoid deleting entries when k and k'
are the same as j or j'.

Example 3.14 Impedance Matching with Ideal Transformer

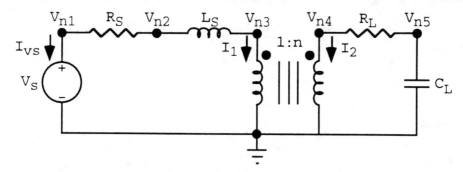

Figure 3.52 Circuit with Ideal Transformer [V_S = 440 V rms, R_S = 1 Ω, R_L =
       60 Ω, L_S = 1/(40π) H, C_L = 625/(3π) μF].

For the circuit shown in Fig. 3.52, where the ideal transformer couples a load
impedance to a source

a) Compute the average power P into the primary side of the ideal transformer
as a function of frequency for values of frequency ranging from 6 Hz to
600 kHz and with values of the turns ratio equal to 2, 4, 6, and 8. Plot the
average power vs. the logarithm of the frequency with turns ratio n as a
parameter.

b) With frequency f equal to 60 Hz, plot the average power as a function of the turns ratio n. Vary n from 2 to 8 in steps of 0.1. Plot power vs. turns ratio, and use the ginput command to measure the maximum power P_{max} and the turns ratio n_{max} that gives this maximum power.

Solution

a) The MNA equation for the circuit in Fig. 3.52 is

$$
\begin{bmatrix}
G_s & -G_s & 0 & 0 & 0 & 1 & 0 & 0 \\
-G_s & G_s & 0 & 0 & 0 & 0 & 1 & 0 \\
0 & 0 & 0 & 0 & 0 & 0 & -1 & 0 \\
0 & 0 & 0 & G_L & -G_L & 0 & 0 & -n \\
0 & 0 & 0 & -G_L & sC_L+G_L & 0 & 0 & 1 \\
1 & 0 & 0 & 0 & 0 & 0 & 0 & 0 \\
0 & 1 & -1 & 0 & 0 & 0 & -sL_s & 0 \\
0 & 0 & 0 & -n & 1 & 0 & 0 & 0
\end{bmatrix}
*
\begin{bmatrix}
V_{n1} \\
V_{n2} \\
V_{n3} \\
V_{n4} \\
V_{n5} \\
I_{vs} \\
I_{ls} \\
I_2
\end{bmatrix}
=
\begin{bmatrix}
0 \\
0 \\
0 \\
0 \\
0 \\
V_s \\
0 \\
0
\end{bmatrix}
$$

The M-file in Fig. 3.53 implements the computation of power P as a function of frequency with turns ratio n as parameter. A few iterations of the MATLAB computations, varying the coordinates in the text commands, results in the plot of Fig. 3.54. The curves show that the power maximizes somewhere between n = 6 and n = 8 at 60 Hz. Also, for a fixed value n, each power curve maximizes at a different frequency.

```
% Example 3.14 - Part a - Impedance Matching with Ideal Transformer
clear
Vs=120;              GS=1;          GL=1/60;
LS=1/(40*pi);        CL=625*1e-6/(3*pi);
G=nmAcc(1,2,GS,zeros(8));        G=nmAcc(4,5,GL,G);
G=mnaVSrc(6,1,0,G);              G=mnaVSrc(7,2,3,G);
C=zeros(8);          C(5,5)=CL;   C(7,7)=-LS;
W=[0 0 0 0 0 Vs 0 0]';
nnp=4;               nnf=21;      n=linspace(2,8,nnp);
f=60*logspace(-1,1,nnf);         s=j*2*pi*f;
for ii=1:nnp
  Gw=mnaIdX1(8,3,0,4,0,n(ii),G);
  for jj=1:nnf
    T=s(jj)*C+Gw;,  X=T\W;
    P(ii,jj)=real(X(3)*conj(-n(ii)*X(8)))/1e3;
  end
end
semilogx(f,P(1,:),'-',f,P(2,:),'--',f,P(3,:),':',f,P(4,:),'-.')
xlabel('Frequency, Hz'), ylabel('Power, kW')
text(40,0.2,'n=2'), text(30,1.2,'n=4')
text(23,2,'n=6'), text(10,2,'n=8'), grid
```

Figure 3.53 M-File to Determine Power vs. Frequency with Turns Ratio as
    Parameter.

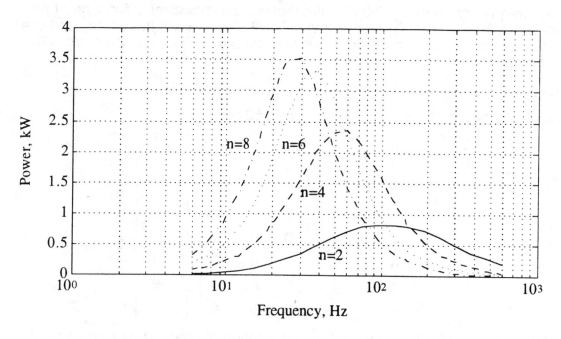

Figure 3.54 Power to Ideal Transformer vs. Frequency with n as Parameter.

b) To determine the value of n that gives maximum power at 60 Hz, use the M-file shown in Fig. 3.55 to create the plot in Fig. 3.56. Then, type

> [nMax,Pmax]=ginput(1) <CR>

in the command window. Position the cross-hair cursor at the peak of the curve and click the mouse button to see

```
% Example 3.14 - Part b - Impedance Matching with Ideal Transformer
clear
Vs=120;       GS=1; GL=1/60;     LS=1/(40*pi);       CL=625*1e-6/(3*pi);
G=nmAcc(1,2,GS,zeros(8));       G=nmAcc(4,5,GL,G);
G=mnaVSrc(6,1,0,G);       G=mnaVSrc(7,2,3,G);
C=zeros(8); C(5,5)=CL;   C(7,7)=-LS;
W=[0 0 0 0 0 Vs 0 0]';
nnp=61;
n=linspace(2,8,nnp);
f=60;        s=j*2*pi*f;
for ii=1:nnp
  Gw=mnaIdX1(8,3,0,4,0,n(ii),G);
  T=s*C+Gw;,  X=T\W;
  P(ii)=real(X(3)*conj(-n(ii)*X(8)))/1e3;
end
plot(n,P)
xlabel('Turns Ratio n'), ylabel('Power, kW')
grid
```

Fig 3.55 M-File to Determine n_{max} and P_{max} at f = 60 Hz.

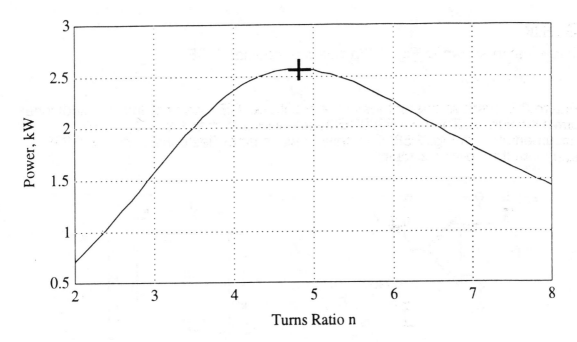

Figure 3.56 Power vs. Turns Ratio at 60 Hz.

```
nMax  =  4.7817
Pmax  =  2.5735
```

According to theory, the maximum power occurs when the reflected magnitude of the load impedance equals the magnitude of the source impedance. Since

$$X_s = \omega L_s = (2\pi \times 60)\frac{1}{40\pi} = 3\ \Omega$$

and

$$X_L = \frac{1}{\omega C_L} = \frac{1}{(2\pi \times 60)\dfrac{625}{3\pi} \times 10^{-6}} = 40\ \Omega$$

then the turns ration n_{max} for maximum power transfer to the load is

$$n_{max} = \sqrt[4]{\frac{R_L^2 + X_L^2}{R_s^2 + X_s^2}} = \sqrt[4]{\frac{80^2 + 40^2}{1^2 + 3^2}} = \sqrt[4]{520} = 4.775$$

The MATLAB experimental and theoretical values of n_{max} agree within 0.15 %. With this value of the turns ratio, the maximum power that the load absorbs is

$$P_{max} = \frac{|V_{srms}|^2}{n_{max}^2}\frac{\mathrm{Re}\{Z_L\}}{\left|Z_s + \dfrac{Z_L}{n_{max}^2}\right|^2} = \frac{120^2}{\sqrt{520}}\frac{\mathrm{Re}\{60 - j40\}}{\left|1 + j3 + \dfrac{60 - j40}{\sqrt{520}}\right|^2} = 2.571\ \text{kW}$$

and the MATLAB value is within 0.1% of this value.

Op Amp

The op amp shown in Fig. 3.57a has one additional CE

$$V_j - V_{j'} = 0 \qquad\qquad\qquad (3.57)$$

Figure 3.57b shows the entries in the additional CE rows and the two additional current columns. The new MATLAB function `mnaOA`, has the `help` documentation in Fig. 3.58. This new function simplifies construction of MNA equations for op-amp circuits.

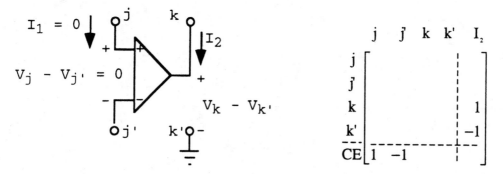

a) Op Amp.   b) MNA Matrix Representation.

Figure 3.57 MNA Representation of Op Amp.

```
amat=mnaOA(m,j,jp,k,bmat)
   Accumulation of entries for an ideal op amp onto bmat.
     m is the index into the added row.
     j and jp are the noninverting and the inverting op-amp
       input nodes.
     k is the output node.
```

Figure 3.58 mnaOA Help.

Example 3.15 Tow-Thomas Quadratic Bandpass Circuit

Use MATLAB to construct the Bode magnitude and phase plot of output voltage V_o for the Tow-Thomas quadratic bandpass filter circuit shown in Fig. 3.59. Vary ω over two decades, ranging from 0.1 krad/s to 10 krad/s and using 20 points per decade.

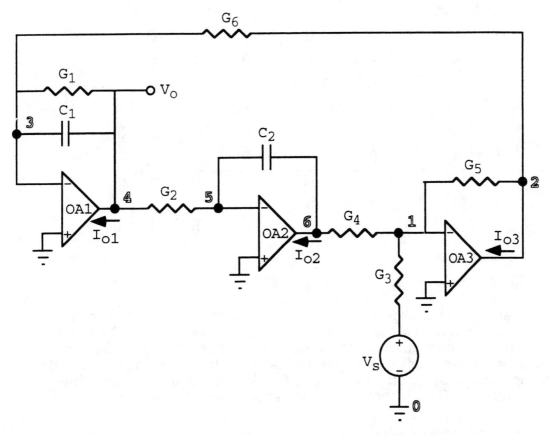

Figure 3.59 Tow-Thomas Biquad Filter Circuit as a Bandpass Filter (G_1 = G/Q = 50 µS, G_2 = G_6 = G = 0.1 mS, G_3 = A_oQ/g = 0.1 mS, G_4 = G_5 = g = 0.1 mS, C_1 = C_2 = C = 0.1 µF).

Solution

The MNA equation for the circuit in Fig. 3.59 is

$$\{sC + G\}X = W$$

where capacitance matrix C is

$$C = \begin{bmatrix} 0 & 0 & 0 & 0 & 0 & 0 & 0 & 0 & 0 \\ 0 & 0 & 0 & 0 & 0 & 0 & 0 & 0 & 0 \\ 0 & 0 & C_1 & -C_1 & 0 & 0 & 0 & 0 & 0 \\ 0 & 0 & -C_1 & C_1 & 0 & 0 & 0 & 0 & 0 \\ 0 & 0 & 0 & 0 & C_2 & -C_2 & 0 & 0 & 0 \\ 0 & 0 & 0 & 0 & -C_2 & C_2 & 0 & 0 & 0 \\ 0 & 0 & 0 & 0 & 0 & 0 & 0 & 0 & 0 \\ 0 & 0 & 0 & 0 & 0 & 0 & 0 & 0 & 0 \\ 0 & 0 & 0 & 0 & 0 & 0 & 0 & 0 & 0 \end{bmatrix}$$

conductance matrix $\mathcal{G}$ is

$$
\mathcal{G} = \left[\begin{array}{cccccc:ccc}
G_3+G_4+G_5 & -G_5 & 0 & 0 & 0 & -G_4 & 0 & 0 & 0 \\
-G_5 & G_5+G_6 & -G_6 & 0 & 0 & 0 & 0 & 0 & 1 \\
0 & -G_6 & G_1+G_6 & -G_1 & 0 & 0 & 0 & 0 & 0 \\
0 & 0 & -G_1 & G_1+G_2 & -G_2 & 0 & 1 & 0 & 0 \\
0 & 0 & 0 & -G_2 & G_2 & 0 & 0 & 0 & 0 \\
-G_4 & 0 & 0 & 0 & 0 & G_4 & 0 & 1 & 0 \\ \hdashline
0 & 0 & -1 & 0 & 0 & 0 & 0 & 0 & 0 \\
0 & 0 & 0 & 0 & -1 & 0 & 0 & 0 & 0 \\
-1 & 0 & 0 & 0 & 0 & 0 & 0 & 0 & 0
\end{array}\right]
$$

and the excitation vector $\mathcal{W}$ is

$$
\mathcal{W} = \begin{bmatrix} G_3 V_s & 0 & 0 & 0 & 0 & 0 & 0 & 0 & 0 \end{bmatrix}^t
$$

In this example, the voltage source V_s and conductance G_3 appear in Norton equivalent form to reduce the number of equations from 11 to 9. The M-file to obtain the Bode magnitude plot for this circuit is shown in Fig. 3.60a. Figure 3.60b shows a modification of this M-file that gives the phase plot. The Bode magnitude and phase plots are in Fig. 3.61. Analysis of this circuit with $G_2 = G_3 = G$ and $G_4 = G_5 = g$ shows that the center frequency ω_o is given by

$$
\omega_o = \frac{G}{C} = \frac{0.1}{0.1} = 1 \text{ krad/s}
$$

The quality factor Q is

$$
Q = \frac{G}{G_1} = \frac{0.1}{0.05} = 2
$$

The half-power and 45° phase shift frequencies are

$$
\omega_H = \omega_{45} = \left\{ \sqrt{1+\left(\frac{1}{2Q}\right)^2} \pm \frac{1}{2Q} \right\} \omega_o = \left\{ \sqrt{1+\left(\frac{1}{2\times 2}\right)^2} \pm \frac{1}{2\times 2} \right\} \times 1
$$

$$
= 0.7808, 1.2808 \text{ krad/s}
$$

The center frequency gain A_o is

$$
A_o = Q\frac{G}{g} = 2\frac{0.1}{0.1} = 2 \qquad (6.02 \text{ dB})
$$

You can use MATLAB's `ginput` function with these plots to confirm these values.

```
% Example 3.15 - Part a - Tow-Thomas Quadratic Bandpass Circuit
%      Units are V, mA, ms, krad/s, kOhm, and uF.
clear
Vs=1;
G1=0.05;      G2=0.1;       G3=0.1;
G4=0.1;       G5=0.1;       G6=0.1;
C1=0.1;       C2=0.1;
C=nmAcc(3,4,C1,zeros(9));        C=nmAcc(5,6,C2,C);
G=nmAcc(3,4,G1,zeros(9));        G=nmAcc(4,5,G2,G);
G=nmAcc(1,0,G3,G);               G=nmAcc(6,1,G4,G);
G=nmAcc(1,2,G5,G);               G=nmAcc(2,3,G6,G);
G=mnaOA(7,0,3,4,G);              G=mnaOA(8,0,5,6,G);
G=mnaOA(9,0,1,2,G);
W=[G3*Vs 0 0 0 0 0 0 0 0]';
nnf=41;
w=logspace(-1,1,nnf);            s=j*w;
for ii=1:nnf
  T=s(ii)*C+G;
  X=T\W;
  Vo(ii)=X(4);
end
semilogx(w,20*log10(Vo))
xlabel('Frequency, kRad/s')
ylabel('Magnitude of Vo, dB')
grid
```

a) M-File for Bode Magnitude Plot.

```
semilogx(w,180*angle(Vo)/pi)
xlabel('Frequency, kRad/s')
ylabel('Phase of Vo, Degrees')
```

b) Modification for Bode Phase Plot.

Figure 3.60 M-Files for Example 3.15.

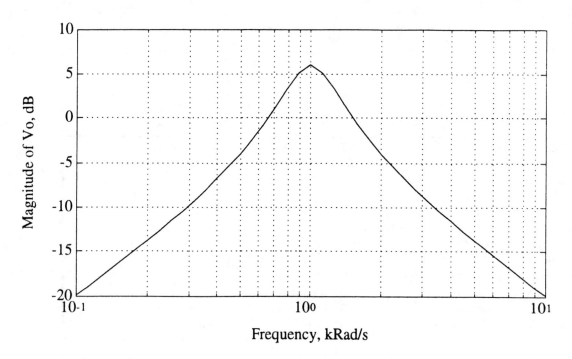

a) Magnitude Plot.

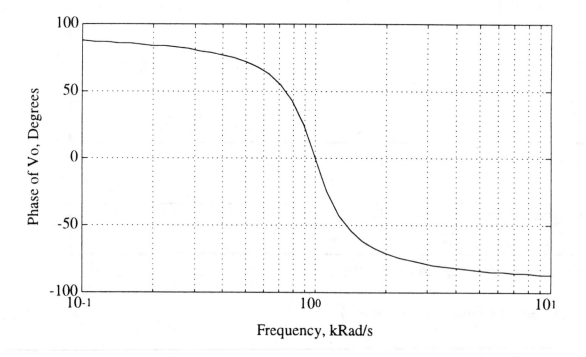

b) Phase Plot

Figure 3.61 Bode Plot for Tow-Thomas Quadratic Bandpass Filter.

Chapter 4

Dynamic Analysis

This chapter provides examples of MATLAB commands useful for dynamic analysis of linear electrical circuits. These commands include Bode and Nyquist plotting, evaluation of natural frequencies, numerical evaluation of transient solutions, and transfer-function residue evaluation.

4.1 Bode and Nyquist Plotting

The MATLAB `bode` and `nyquist` commands[1] provide virtually immediate Bode and Nyquist plots from the $\mathcal{A}$, $\mathcal{B}$, $\mathcal{C}$, and $\mathcal{D}$ matrices that appear in the state-variable equation

$$\dot{X} = \mathcal{A} * X + \mathcal{B} * X_s \tag{4.1}$$

and the output equation

$$\mathcal{Y} = C * X + \mathcal{D} * X_s \tag{4.2}$$

In an electrical circuit, the state-variable matrix X is a column vector whose components are the independent inductance currents I_{lj} and independent capacitance voltages V_{ck}. For example, if a circuit has n_L independent inductance currents and n_C independent capacitance voltages, then

$$X = \begin{bmatrix} I_{l1} & I_{l2} & \cdots & I_{ln_L} & \vdots & V_{c1} & V_{c2} & \cdots & V_{cn_C} \end{bmatrix}^t$$

The source vector X_s is a vector that consists of the independent current and voltage sources in the circuit. Response matrix $\mathcal{Y}$ is a vector having voltage or current components of interest. These depend linearly on the state variables and the circuit's independent sources. Matrices $\mathcal{A}$, $\mathcal{B}$, C, and $\mathcal{D}$ depend on the circuit's parameters. In Laplace transform notation the state-variable equations are

$$sX = \mathcal{A}X + \mathcal{B}X_s \tag{4.3}$$

$$\mathcal{Y} = CX + \mathcal{D}X_s \tag{4.4}$$

where initial-condition sources exist as components of source matrix X_s.

With $\mathcal{A}$, $\mathcal{B}$, C, and $\mathcal{D}$ known, use any of the following forms

```
bode(a,b,c,d)
bode(a,b,c,d,i)
bode(a,b,c,d,i,w)
```

Of course, parameters a, b, c, and d are the matrices $\mathcal{A}$, $\mathcal{B}$, C, and $\mathcal{D}$. Integer i identifies the single component of X_S to use as the circuit's excitation. Array w identifies the radian frequency values that MATLAB uses to calculate the

---

[1] The `bode` and `nyquist` commands exist in the Signals and Systems Toolbox of The Student Edition of MATLAB 3.5. If you use MATLAB 4.X, you will find these functions in the optional Control Systems Toolbox.

response. If you omit w, then MATLAB selects the frequency range automatically and uses more frequency values when the response magnitude and angle vary rapidly. The two additional forms

```
bode(num,den)
bode(num,den,w)
```

provide Bode plots for transfer functions, where num and den give the array of coefficients of the numerator and denominator polynomials in descending powers of s. For example, if the numerator polynomial is

$2s^2+5s+3$

then use

```
»num=[2 5 3]
```

If you wish to inspect the magnitude and phase values that MATLAB computes, use an assignment, such as

```
»[Mag,Phase,w]=bode(a,b,c,d)
```

Mag contains the magnitude, not the dB, values of response vector $\mathcal{Y}$. Phase has the angles in degrees and column vector w the frequency values. Mag and Phase have as many columns as there are components of the response vector $\mathcal{Y}$. The number of rows in Mag and Phase correspond to the number of values in w. The assignment form does not do a screen plot. If you wish to make a Bode plot with the frequency in Hz rather than in rad/s, you can use the Mag and Phase matrices, scaling w by 2π. Then use loglog, convert Mag to dB with

```
»dBVal=20*log10(Mag);
```

and add appropriate labels or titles, using xlabel, ylable, and title. The subplot command is useful here to see multiple plots in the same window. The syntax of this command in MATLAB 3.5 is

```
»subplot(mnp)
```

where m and n give the number of rows and columns in a grid of subplots. Index p indicates which subplot is active, counting row by row from top to bottom. The values of m or n may not exceed two, and index p can be no greater than m×n. For example

```
»subplot(223)
```

activates the subplot in the second row, first column of a 2×2 plot. In MATLAB 4.X, subplot takes integers m, n, and p as separate arguments. Although MATLAB 4.X still supports the old version of subplot, using

```
»subplot(m,n,p)
```

is better, since future versions may discontinue support of the old syntax.

The nyquist command uses the same arguments as bode. In the assignment form of nyquist, the return arguments are the real part, the imaginary part of the response, and the frequency w.

Examples 4.1 and 4.2 show how to use functions bode and nyquist. In particular, these examples show how to cast the circuit equations into state-variable form, identifying matrices $\mathcal{A}$, $\mathcal{B}$, $\mathcal{C}$, and $\mathcal{D}$ for use by bode and nyquist.

Example 4.1 Bode Plot of Tow-Thomas Quadratic Bandpass Filter

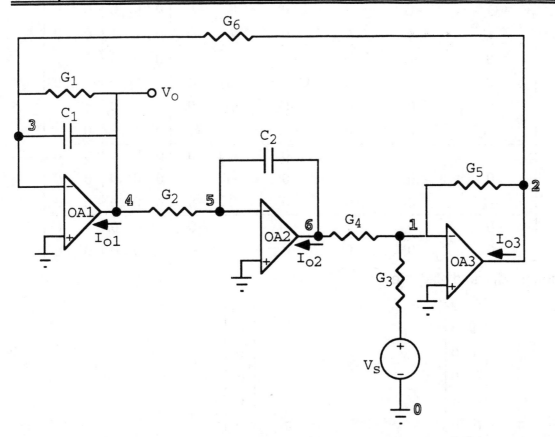

Figure 4.1 Tow-Thomas Biquad Filter Circuit as a Bandpass Filter (G_1 = G/Q = 50 μS, G_2 = G_6 = G = 100 μS, G_3 = A_oQ/g = 100 μS, G_4 = G_5 = g = 100 μS, C_1 = C_2 = C = 100 nF).

Figure 4.1 reproduces Fig. 3.58 and shows the Tow-Thomas quadratic bandpass circuit. Produce the Bode plot for this circuit using MATLAB's bode command.

Solution

To construct the four matrices $\mathcal{A}$, $\mathcal{B}$, $\mathcal{C}$, and $\mathcal{D}$ of the state-variable formulation for this circuit, start with the MNA equation

$$G_{mn}\mathcal{V}_{mna} = \mathcal{A}_{mx}X + \mathcal{A}_{ms}X_s$$

Here, the MNA solution vector $\mathcal{V}_{mna}$ is

$$\mathcal{V}_{mna} = \begin{bmatrix} V_{n1} & V_{n2} & V_{n3} & V_{n4} & V_{n5} & V_{n6} \mid I_{o1} & I_{o2} & I_{o3} \mid I_{c1} & I_{c2} \end{bmatrix}^t$$

Consisting of the two independent capacitance voltages, the state-variable vector X is

$$X = \begin{bmatrix} V_{c1} & V_{c2} \end{bmatrix}^t$$

and the independent-source vector X_s is simply $[V_s]$ = V_s.

The circuit's MNA matrix $\mathcal{G}_{mm}$ is

$$\mathcal{G}_{mm} = \left[\begin{array}{cccccc:ccc:cc}
G_3+G_4+G_5 & -G_5 & 0 & 0 & 0 & -G_4 & 0 & 0 & 0 & 0 & 0 \\
-G_5 & G_5+G_6 & -G_6 & 0 & 0 & 0 & 0 & 0 & 1 & 0 & 0 \\
0 & -G_6 & G_1+G_6 & -G_1 & 0 & 0 & 0 & 0 & 0 & 1 & 0 \\
0 & 0 & -G_1 & G_1+G_2 & -G_2 & 0 & 1 & 0 & 0 & -1 & 0 \\
0 & 0 & 0 & -G_2 & G_2 & 0 & 0 & 0 & 0 & 0 & 1 \\
-G_4 & 0 & 0 & 0 & 0 & G_4 & 0 & 1 & 0 & 0 & -1 \\ \hdashline
0 & 0 & -1 & 0 & 0 & 0 & 0 & 0 & 0 & 0 & 0 \\
0 & 0 & 0 & 0 & -1 & 0 & 0 & 0 & 0 & 0 & 0 \\
-1 & 0 & 0 & 0 & 0 & 0 & 0 & 0 & 0 & 0 & 0 \\ \hdashline
0 & 0 & 1 & -1 & 0 & 0 & 0 & 0 & 0 & 0 & 0 \\
0 & 0 & 0 & 0 & 1 & -1 & 0 & 0 & 0 & 0 & 0
\end{array}\right]$$

Matrices $\mathcal{A}_{mx}$ and $\mathcal{A}_{ms}$ are

$$\mathcal{A}_{mx} = \left[\begin{array}{cc}
0 & 0 \\
0 & 0 \\
0 & 0 \\
0 & 0 \\
0 & 0 \\
0 & 0 \\ \hdashline
0 & 0 \\
0 & 0 \\
0 & 0 \\ \hdashline
1 & 0 \\
0 & 1
\end{array}\right]
\qquad
\mathcal{A}_{ms} = \left[\begin{array}{c}
G_3 \\
0 \\
0 \\
0 \\
0 \\
0 \\ \hdashline
0 \\
0 \\
0 \\ \hdashline
0 \\
0
\end{array}\right]$$

These matrices account for terms in the MNA equations due to V_{c1}, V_{c2}, and V_s. The capacitance-current matrix is

$$\begin{bmatrix} I_{c1} \\ I_{c2} \end{bmatrix} = \mathcal{V}_{mm}([10,11],:) = j\omega\begin{bmatrix} C_1 & 0 \\ 0 & C_2 \end{bmatrix} * \begin{bmatrix} V_{c1} \\ V_{c2} \end{bmatrix} = j\omega C_{xx} * X$$

Then

$$j\omega X = C_{xx}^{-1}\left\{\left(\mathcal{G}_{mm}^{-1}\mathcal{A}_{mx}\right)X + \mathcal{G}_{mm}^{-1}\mathcal{A}_{ms}X_s\right\}_{(10:11)} = \mathcal{A}X + \mathcal{B}X_s$$

Here, state-variable matrix $\mathcal{A}$ is

$$\mathcal{A} = C_{xx}^{-1}\left\{\mathcal{G}_{mm}^{-1}\mathcal{A}_{mx}\right\}_{(10:11)}$$

and state-variable matrix $\mathcal{B}$ is

$$\mathcal{B} = C_{xx}^{-1}\left\{\mathcal{G}_{mm}^{-1}\mathcal{A}_{ms}\right\}_{(10:11)}$$

In these equations the subscript notation (10:11) denotes the 10-th and 11-th rows of the matrix. Since the output of interest is $V_o = V_4$, the output matrix $\mathcal{Y}$ is

$$\mathcal{Y} = V_o = \mathcal{V}_m(4) = \left(\mathcal{G}_{mm}^{-1}\mathcal{A}_{mx}\right)_{(4)}\mathcal{X} + \left(\mathcal{G}_{mm}^{-1}\mathcal{A}_{ms}\right)_{(4)}\mathcal{X}_s = C\mathcal{X} + D\mathcal{X}_s$$

Then

$$C = \left\{\mathcal{G}_{mm}^{-1}\mathcal{A}_{mx}\right\}_{(4)}$$

$$\mathcal{D} = \left\{\mathcal{G}_{mm}^{-1}\mathcal{A}_{ms}\right\}_{(4)}$$

The M-file in Fig. 4.2 produces the Bode plot in Fig. 4.3, showing the output voltage V_o as a function of frequency ω. To obtain the Nyquist plot in Fig. 4.4, replace the bode function in the last line of Fig. 4.2 with the nyquist function.

```
% Example 4.1 - Bode Plot of Tow Thomas Quadratic Bandpass Filter
%     Units are V, mA, ms, krad/s, kOhm, and uF.
clear, clg
Vs=1;
G1=0.05;    G2=0.1;     G3=0.1;
G4=0.1;     G5=0.1;     G6=0.1;
C1=0.1;, C2=0.1;
Cxx=nmAcc(1,0,C1,zeros(2));
Cxx=nmAcc(2,0,C2,Cxx);
Gmm=nmAcc(3,4,G1,zeros(11));
Gmm=nmAcc(4,5,G2,Gmm);
Gmm=nmAcc(1,0,G3,Gmm);
Gmm=nmAcc(6,1,G4,Gmm);
Gmm=nmAcc(1,2,G5,Gmm);
Gmm=nmAcc(2,3,G6,Gmm);
Gmm=mnaOA(7,0,3,4,Gmm);
Gmm=mnaOA(8,0,5,6,Gmm);
Gmm=mnaOA(9,0,1,2,Gmm);
Gmm=mnaVSrc(10,3,4,Gmm);
Gmm=mnaVSrc(11,5,6,Gmm);
Ams=SrcAcc(0,1,G3,zeros(11,1));
Amx=zeros(11,2);
Amx(10,1)=1;
Amx(11,2)=1;
Gmx=Gmm\Amx;
Gms=Gmm\Ams;
% The A, B, C, And D matrices are
A=Cxx\Gmx([10,11],:);
B=Cxx\Gms([10,11],:);
C=Gmx(4,:);
D=Gms(4,:);
bode(A,B,C,D)
% nyquist(A,B,C,D)
```

Figure 4.2 M-File for Bode Plot of Tow-Thomas Quadratic Bandpass Filter.

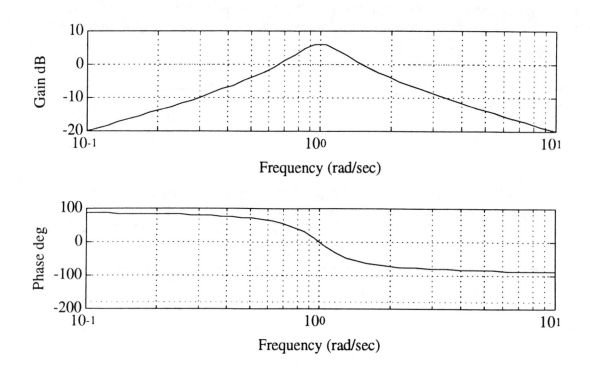

Figure 4.3 Bode Plot of Tow-Thomas Quadratic Bandpass Filter
(Note: Frequency is in krad/s, not rad/s).

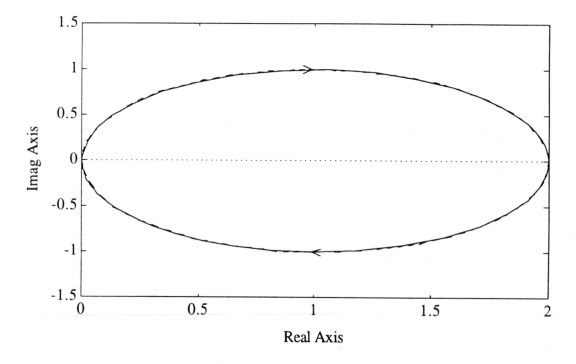

Figure 4.4 Nyquist Plot of Tow-Thomas Quadratic Bandpass Filter.

Example 4.2 Cascode Amplifier

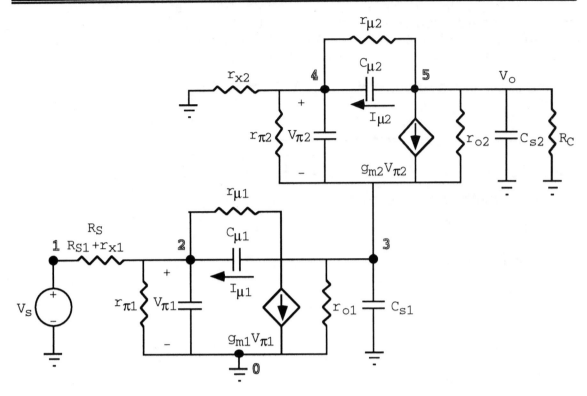

Figure 4.5 Cascode Amplifier (R_{S1} = 1 kΩ, R_C = 10 kΩ, r_{x1} = r_{x2} = 0.5 kΩ, $r_{\pi1}$ = $r_{\pi2}$ = 2.5 kΩ, r_{o1} = r_{o2} = 100 kΩ, $r_{\mu1}$ = $r_{\mu2}$ = 18 MΩ, g_{m1} = g_{m2} = 40 mS, $C_{\pi1}$ = $C_{\pi2}$ = 15 pF, $C_{\mu1}$ = $C_{\mu2}$ = 0.6 pF, C_{s1} = 2 pF, and C_{s2} = 1.5 pF).

For the cascode amplifier circuit shown in Fig. 4.5, plot the Bode magnitude and phase curves.

Solution

This problem is of interest because it involves dependent capacitances, whose voltages are not components of the state-variable solution set. Capacitances $C_{\pi1}$, $C_{\mu1}$, and C_{s1} form one loop. Capacitances C_{s1}, $C_{\pi2}$, $C_{\mu2}$, and C_{s2} form a second loop. We can break these two loops by removing capacitances $C_{\mu1}$ and $C_{\mu2}$ from the circuit. Removal of other capacitance pairs to break these two loops is possible, but the choice here is convenient. The existence of a capacitance, voltage-source loop in a circuit means that one capacitance in the loop can not have an arbitrary voltage value without violating KVL, so one of these capacitances is not an independent energy storage element. The voltage in this dependent capacitance can not be a member of the state-variable solution matrix. Similarly, existence of an inductance, current-source star connection means that one inductance in the star can not have an arbitrary current value without violating KCL, so one of these inductances is not an independent energy storage element. The current in this dependent inductance can not be a member of the state-variable solution matrix. To cast the circuit equations into state-variable form, we have to decide what to do with these dependent energy-storage elements. The solution to this problem is to represent a dependent capacitance as an independent current source and a dependent inductance as an independent voltage source.

For the circuit in this example, write the MNA equations in the form

$$\mathcal{G}_{mm} \mathcal{V}_{mna} = \mathcal{A}_{mc} I_c + \mathcal{A}_{mx} X + \mathcal{A}_{ms} X_s$$

where the MNA solution vector $\mathcal{V}_{mna}$ is

$$\mathcal{V}_{mna} = \begin{bmatrix} V_{n1} & V_{n2} & V_{n3} & V_{n4} & V_{n5} & | & I_{vs} & | & I_{x1} & I_{s1} & I_{x2} & I_{s2} \end{bmatrix}^t$$

The dependent capacitance current vector I_c is

$$I_c = \begin{bmatrix} I_{\mu 1} & I_{\mu 2} \end{bmatrix}^t$$

State-variable matrix X is

$$X = \begin{bmatrix} V_{\pi 1} & V_{s1} & V_{\pi 2} & V_{s2} \end{bmatrix}^t$$

and the excitation vector X_s is

$$X_s = \begin{bmatrix} V_s \end{bmatrix} = V_s$$

The MNA matrix $\mathcal{G}_{mm}$ is

$$\mathcal{G}_{mm} = \begin{bmatrix} \mathcal{G}_{nn} & | & \mathcal{A}_{mx} \\ \hline \mathcal{A}_{ce} & | & 0_{5 \times 5} \end{bmatrix}$$

with

$$\mathcal{G}_{nn} = \begin{bmatrix} G_S & -G_S & 0 & 0 & 0 \\ -G_S & G_S + g_{\pi 1} + g_{\mu 1} & -g_{\mu 1} & 0 & 0 \\ 0 & g_{m1} - g_{\mu 1} & g_{m2} + g_{o1} + g_{\mu 1} + g_{\pi 2} + g_{o2} & -g_{m2} - g_{\pi 2} & -g_{o2} \\ 0 & 0 & -g_{\pi 2} & g_{\pi 2} + g_{x2} + g_{\mu 2} & -g_{\mu 2} \\ 0 & 0 & -g_{m2} - g_{o2} & g_{m2} - g_{\mu 2} & G_C + g_{o2} + g_{\mu 2} \end{bmatrix}$$

Submatrices $\mathcal{A}_{mi}$ and $\mathcal{A}_{ce}$ are

$$\mathcal{A}_{mi} = \begin{bmatrix} 1 & | & 0 & 0 & 0 & 0 \\ \hline 0 & | & 1 & 0 & 0 & 0 \\ 0 & | & 0 & 1 & -1 & 0 \\ 0 & | & 0 & 0 & 1 & 0 \\ 0 & | & 0 & 0 & 0 & 1 \end{bmatrix} \qquad \mathcal{A}_{ce} = \begin{bmatrix} 1 & 0 & 0 & 0 & 0 \\ \hline 0 & 1 & 0 & 0 & 0 \\ 0 & 0 & 1 & 0 & 0 \\ 0 & 0 & -1 & 1 & 0 \\ 0 & 0 & 0 & 0 & 1 \end{bmatrix} = \mathcal{A}_{mi}^t$$

Matrices $\mathcal{A}_{mc}$, $\mathcal{A}_{mx}$, and $\mathcal{A}_{ms}$ are

$$\mathcal{A}_{mc} = \begin{bmatrix} 0 & 0 \\ 1 & 0 \\ -1 & 0 \\ 0 & 1 \\ 0 & -1 \\ \hline 0_{5\times2} \end{bmatrix} \qquad \mathcal{A}_{mx} = \begin{bmatrix} 0_{5\times4} \\ \hline 0 & 0 & 0 & 0 \\ 1 & 0 & 0 & 0 \\ 0 & 1 & 0 & 0 \\ 0 & 0 & 1 & 0 \\ 0 & 0 & 0 & 1 \end{bmatrix} \qquad \mathcal{A}_{ms} = \begin{bmatrix} 0_{5\times1} \\ \hline 1 \\ 0 \\ 0 \\ 0 \\ 0 \end{bmatrix}$$

The dependent capacitance current matrix I_C depends only on the state-variable matrix and is

$$I_c = j\omega C_{cx} X = j\omega \begin{bmatrix} -C_{\mu1} & C_{\mu1} & 0 & 0 \\ 0 & -C_{\mu2} & -C_{\mu2} & C_{\mu2} \end{bmatrix} * \begin{bmatrix} V_{x1} \\ V_{s1} \\ V_{x2} \\ V_{s2} \end{bmatrix}$$

This equation may have a second term proportional to the source vector X_s if any loops include voltage sources. In any case, the dependent capacitance voltages depend solely on the voltages of the other voltage sources and capacitances in their loop. To extract the state-variable equations from these equations, write the vector of independent capacitance currents in terms of the state-variable vector using both the capacitance constitutive equation and the last four rows of the solution of the MNA equation

$$\begin{bmatrix} I_{x1} \\ I_{s1} \\ I_{x2} \\ I_{s2} \end{bmatrix} = V_{mns}(7:10,:) = j\omega \begin{bmatrix} C_{x1} & 0 & 0 & 0 \\ 0 & C_{s1} & 0 & 0 \\ 0 & 0 & C_{x2} & 0 \\ 0 & 0 & 0 & C_{s2} \end{bmatrix} * \begin{bmatrix} V_{x1} \\ V_{s1} \\ V_{x2} \\ V_{s2} \end{bmatrix} = j\omega C_{xx} X$$

$$= j\omega \left[G_{mm}^{-1} \mathcal{A}_{mc} C_{cx} \right]_{(7:10)} X + \left[G_{mm}^{-1} \mathcal{A}_{mx} \right]_{(7:10)} X + \left[G_{mm}^{-1} \mathcal{A}_{ms} \right]_{(7:10)} X_s$$

The subscript notation (7:10) means that the matrix consists of the 7-th through the 10-th rows and all columns of the reference matrix. Now, define the effective capacitance matrix C_t

$$C_t = C_{xx} - \left\{ G_{mm}^{-1} \mathcal{A}_{mc} C_{cx} \right\}_{(7:10)}$$

Then the state-variable equation is

$$j\omega X = \mathcal{A}X + \mathcal{B}X_s = C_t^{-1} \left\{ G_{mm}^{-1} \mathcal{A}_{mx} \right\}_{(7:10)} X + C_t^{-1} \left\{ G_{mm}^{-1} \mathcal{A}_{ms} \right\}_{(7:10)} X_s$$

and the output equation is

$$\mathcal{Y} = [V_o] = CX + \mathcal{D}X_s$$

$$= [0 \quad 0 \quad 0 \quad 1] \begin{bmatrix} V_{x1} \\ V_{s1} \\ V_{x2} \\ V_{s2} \end{bmatrix} + [0][V_s]$$

```
%   Example 4.2 Cascode Amplifier
%      Solution in State-Variable Form
%      Units are V, A, s, rad/s or Hz, Ohm, S, and F
clear, clg
GS=1/(1E3+500);     GC=1/1E4;
gx2=1/500;
gm1=0.04;           gm2=0.04;
gpi1=1/2.5E3;       gpi2=1/2.5E3;
go1=1/1E5;          go2=1/1E5;
gmu1=1/18E6;        gmu2=1/18E6;
cpi1=15E-12;        cpi2=15E-12;
cmu1=0.6E-12;       cmu2=0.6E-12;
cs1=2E-12;          cs2=1.5E-12;
Gmm=nmAcc(1,2,GS,zeros(6));
Gmm=nmAcc(2,0,gpi1,Gmm);
Gmm=nmAcc(3,2,gmu1,Gmm);
Gmm=nmAcc(3,0,go1,Gmm);
Gmm=nmAcc(4,0,gx2,Gmm);
Gmm=nmAcc(4,3,gpi2,Gmm);
Gmm=nmAcc(5,4,gmu2,Gmm);
Gmm=nmAcc(5,3,go2,Gmm);
Gmm=nmAcc(5,0,GC,Gmm);
Gmm=gmAcc(2,0,3,0,gm1,Gmm);
Gmm=gmAcc(4,3,5,3,gm2,Gmm);
Gmm=mnaVSrc(6,1,0,Gmm);
Gmm=mnaVSrc(7,2,0,Gmm);
Gmm=mnaVSrc(8,3,0,Gmm);
Gmm=mnaVSrc(9,4,3,Gmm);
Gmm=mnaVSrc(10,5,0,Gmm);
Amc=zeros(10,2);
Amc(:,1)=srcAcc(3,2,1,Amc(:,1));
Amc(:,2)=srcAcc(5,4,1,Amc(:,2));
Amx=zeros(10,4);                Amx(7:10,:)=eye(4);
Ams=zeros(10,1);                Ams(6,1)=1;
Ccx=[cmu1*[-1 1 0 0]; cmu2*[0 -1 -1 1]];
Ccx=Gmm\(Amc*Ccx);             Ccx=Ccx(7:10,:);
Cxx(1,1)=cpi1; Cxx(2,2)=cs1;   Cxx(3,3)=cpi2; Cxx(4,4)=cs2;
Ct=Cxx-Ccx;
A=Gmm\Amx;   A=A(7:10,:);            A=Ct\A;
B=Gmm\Ams;   B=B(7:10,:);            B=Ct\B;
C=[0 0 0 1];
D=0;
bode(A,B,C,D)
```

Figure 4.6 M-File to Solve Example 4.2.

Running the M-file in Fig. 4.6 gives the Bode plot in Fig. 4.7. The bode function automatically plots the response for frequencies ranging from 10^6 rad/s to 10^{10} rad/s. However, this circuit model is not valid beyond about 10^9 rad/s. Specifying frequency array w with

        w=logspace(6,8,41)

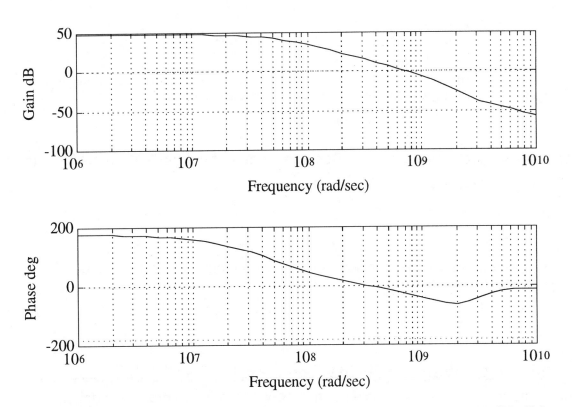

Figure 4.7 Bode Plot for Cascode Amplifier (Note: Frequency is in Grad/s).

and using

```
bode(a,b,c,d,i,w)
```

limits the plot to a meaningful frequency range.

4.2 Natural   Frequencies

With MATLAB, you can easily evaluate natural frequencies or eigenvalues of a circuit using the `eig` function. `eig` has the following forms

```
»eig(A)
»eig(A,'nobalance')
»eig(A,B)
```

Each of these returns the eigenvalue or natural frequencies, which are the values of λ that satisfy the equation

$$A * X = \lambda X$$

or, in the case of the third form

$$A * X = \lambda B * X$$

If you wish to store the eigenvalues in a variable for subsequent calculations, use

```
»[Vector,Diagonal]=eig(A)
»[Vector,Diagonal]=eig(A,'nobalance')
»[Vector,Diagonal]=eig(A,B)
```

The columns of matrix `Vector` are the eigenvectors and the diagonal values of `Diagonal` are the eigenvalues. To recover these as an array use

```
»eig_Vals=diag(Diagonal);
```

For electrical circuit problems, the homogeneous form of the MNA equations are

$$\{sC+G\}*X=0 \qquad\qquad G*X=-sC$$

so use

```
»eig(G,-C)
```

to evaluate the natural frequencies. Alternately, identify the state-variable equations and use matrix $\mathcal{A}$, as in

```
»eig(A)
```

Example 4.3 Natural Frequencies of Tow-Thomas Circuit

Determine the natural frequencies of the Tow-Thomas quadratic bandpass circuit of Fig. 4.1

Solution

Run the M-File of Example 4.1 to generate matrix $\mathcal{A}$ of the state-variable equations for the Tow-Thomas circuit. Then enter

```
»eig(A)
```

to obtain the natural frequencies

```
ans =
-0.2500 + 0.9682i
-0.2500 - 0.9682i
```

Since the center frequency ω_o for the Tow-Thomas circuit is

$$\omega_o = \frac{G}{C} = \frac{0.1}{0.1} = 1\,\text{krad}/s$$

you can compute the natural frequencies using

$$s_P = \omega_o\left\{-\frac{1}{2Q}\pm j\sqrt{1-\left(\frac{1}{2Q}\right)^2}\right\} = 1\left\{-\frac{1}{2\times2}\pm j\sqrt{1-\left(\frac{1}{2\times2}\right)^2}\right\}$$

$$= -0.25 \pm j0.9682\,\text{krad}/s$$

Now, run M-file Ex_3_15.M and enter

```
»eig(G,-C)
```

to have MATLAB compute the natural frequencies from the capacitance and the conductance matrix for the MNA equations. Now the solution is

```
ans =
NaN +       NaNi
NaN +       NaNi
NaN +       NaNi
NaN +       NaNi
NaN +       NaNi
NaN +       NaNi
NaN +       NaNi
-0.2500 + 0.9682i
-0.2500 - 0.9682i
```

The NaN values mean "Not a Number." Although there are nine NMA equations for Example 3.15, we expect two natural frequencies, because C_1 and C_2 are the two energy storage elements in the circuit. MATLAB finds these two values here.

Example 4.4 Cascode Amplifier

For the cascode amplifier circuit shown in Fig. 4.5, determine the natural frequencies. Use these to estimate the half-power frequency as a guide to selecting a frequency range for a Bode plot of the output voltage V_o.

Solution

Run the Example 4.2 M-file. Then type

```
eig(A)
```

to compute the natural frequencies for the cascode amplifier circuit. Running this M-file gives the natural frequencies in Grad/s

```
ans =
   1.0e+09 *
  -1.1732 + 0.9151i
  -1.1732 - 0.9151i
  -0.0466 + 0.0203i
  -0.0466 - 0.0203i
```

Figure 4.8 shows an s-plane plot of these poles. Since the radial distance of the two poles nearest to the origin is approximately 50 Mrad/s, we expect the half-power frequency to be approximately 5 MHz. The poles at -1.17 ± j9.15 Grad/s have little effect on the half-power frequency, because their vectors to s = jω with ω in the vicinity of the half-power frequency are virtually the same length as when s = j0. To see the response at frequencies including 5 MHz, sweep the frequency from 100 kHz to 10 MHz. Knowing the circuit's natural frequencies helps in the selection of an appropriate range of frequencies to use in the M-file that generates the Bode plot.

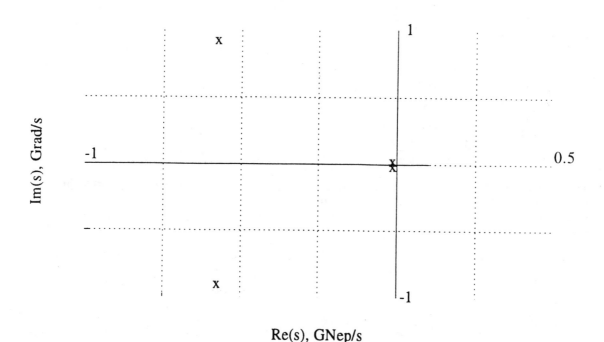

Figure 4.8 S-Plane Plot of Poles of Cascode Amplifier.

4.3 Transient Response

With the functions `ode23` or `ode45` you can calculate the transient response for a circuit. `ode23` uses a second- and third-order Runge-Kutta method for moderate accuracy, while `ode45` uses a fourth- and fifth-order Runge-Kutta method. To use these commands, type

    »[t,x]=ode23('<File_Name>',to,tf,xo,[tol,[trace]])

or

    »[t,x]=ode45('<File_Name>',to,tf,xo,[tol,[trace]])

The first argument <File_Name>, which you type within single quotes, names a function file that returns function f(x,t) in the equation

$$\frac{dx}{dt} = f(x,t)$$

where x is the state variable and t is an array of time values that functions `ode23` and `ode45` return. For Eq. 4.1, the linear state-variable equation, f(x,t) is

$$f(x,t) = \mathcal{A}X + \mathcal{B}X_s$$

The function `StVarEq` in Fig. 4.9 computes the derivatives for any linear circuit problem, where A and B are the state variable equation matrices and rA is the number of rows of matrix A. Variable `to` is the initial time and `tf` is the final time for the solution. `xo` is a column vector that gives the initial values of state-variable x. Optional variable `tol` defaults to 10^{-3} for `ode23` and to 10^{-6} for

```
function xdot = StVarEq(t,x)
%StVarEq(t,x)
%  Define source vector xs, state variable matrices A and B,
%  and rows rA of A as global variables.
%  Comment out the statement "global xs A B" if using
%  The Student Edition of MATLAB 3.5, but define xs, A, B,
%  and rA in your workspace before calling StVarEq.
global xs A B rA
xdot=zeros(rA,1);
xdot=A*x+B*xs;
```

Figure 4.9 Function `StVarEq` to Calculate Solution of Linear State-Variable
      Equations.

`ode45`. Optional parameter `trace` defaults to 0. Any other value causes the time
values, step size, and state values to print as the computation proceeds.

Example 4.5 Third-Order Butterworth Filter

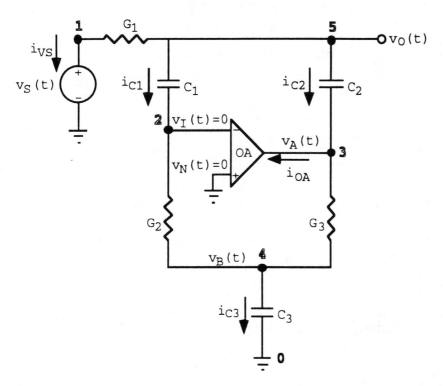

Figure 4.10 Third-Order Butterworth Filter ($R_1 = 1/G_1 = R_2 = 1/G_2 = R_3 = 1/G_3 = 1/G = 1$ kΩ, $C_1 = C_2 = C_3 = C = 1$ μF).

Using MATLAB function `ode23`, plot the output voltage $v_O(t)$ for the third-order
Butterworth filter circuit shown in Fig. 4.10 when the source voltage $v_S(t)$ is a
1-V unit-step source. Superimpose on this plot the theoretical solution

$$v_O(t) = 1 - e^{-t} - \frac{2}{\sqrt{3}} e^{-t/2} \sin\left(\frac{\sqrt{3}}{2} t\right) V$$

Solution

The MNA equation for the third-order Butterworth filter circuit shown in Fig. 4.10 is

$$\mathcal{G}_{MM}\mathcal{V}_M = \mathcal{W} = \mathcal{A}_{MX}\mathcal{X} + \mathcal{A}_{MS}\mathcal{X}_S$$

The MNA solution vector $\mathcal{V}_M$ is

$$\mathcal{V}_M = \begin{bmatrix} v_{N1} & v_{N2} & v_{N3} & v_{N4} & v_{N5} & i_{VS} & i_{OA} & i_{C1} & i_{C2} & i_{C3} \end{bmatrix}^t$$

the state vector $\mathcal{X}$ is

$$\mathcal{X} = \begin{bmatrix} v_{C1} & v_{C2} & v_{C3} \end{bmatrix}^t$$

and the independent-source vector $\mathcal{X}_S$ is

$$\mathcal{X}_S = \begin{bmatrix} v_S \end{bmatrix} = v_S$$

The MNA conductance matrix is

$$\mathcal{G}_{MM} = \begin{bmatrix}
G_1 & 0 & 0 & 0 & -G_1 & 1 & 0 & 0 & 0 & 0 \\
0 & G_2 & 0 & -G_2 & 0 & 0 & 0 & -1 & 0 & 0 \\
0 & 0 & G_3 & -G_3 & 0 & 0 & 1 & 0 & -1 & 0 \\
0 & -G_2 & -G_3 & G_2+G_3 & 0 & 0 & 0 & 0 & 0 & 1 \\
-G_1 & 0 & 0 & 0 & G_1 & 0 & 0 & 1 & 1 & 0 \\
1 & 0 & 0 & 0 & 0 & 0 & 0 & 0 & 0 & 0 \\
0 & -1 & 0 & 0 & 0 & 0 & 0 & 0 & 0 & 0 \\
0 & -1 & 0 & 0 & 1 & 0 & 0 & 0 & 0 & 0 \\
0 & 0 & -1 & 0 & 1 & 0 & 0 & 0 & 0 & 0 \\
0 & 0 & 0 & 1 & 0 & 0 & 0 & 0 & 0 & 0
\end{bmatrix}$$

Matrices $\mathcal{A}_{MX}$ and $\mathcal{A}_{MS}$ are

$$\mathcal{A}_{MX} = \begin{bmatrix}
0 & 0 & 0 \\
0 & 0 & 0 \\
0 & 0 & 0 \\
0 & 0 & 0 \\
0 & 0 & 0 \\
0 & 0 & 0 \\
0 & 0 & 0 \\
1 & 0 & 0 \\
0 & 1 & 0 \\
0 & 0 & 1
\end{bmatrix} \qquad
\mathcal{A}_{MS} = \begin{bmatrix}
0 \\ 0 \\ 0 \\ 0 \\ 0 \\ 1 \\ 0 \\ 0 \\ 0 \\ 0
\end{bmatrix}$$

The capacitance-current matrix is

$$\begin{bmatrix} i_{C1} \\ i_{C2} \\ i_{C3} \end{bmatrix} = \frac{d}{dt}\begin{bmatrix} C_1 & 0 & 0 \\ 0 & C_2 & 0 \\ 0 & 0 & C_3 \end{bmatrix} * \begin{bmatrix} v_{C1} \\ v_{C2} \\ v_{C3} \end{bmatrix} = \frac{d}{dt}C_{XX} * X = \mathcal{V}_M([8:10],:)$$

Then

$$\frac{d}{dt}X = C_{XX}^{-1}\left[\left(\mathcal{G}_{MM}^{-1}\mathcal{A}_{MX}\right)_{(8:10)}X + \left(\mathcal{G}_{MM}^{-1}\mathcal{A}_{MS}\right)_{(8:10)}X_S\right] = \mathcal{A}X + \mathcal{B}X_S$$

where matrix $\mathcal{A}$ is

$$\mathcal{A} = C_{XX}^{-1}\left(\mathcal{G}_{MM}^{-1}\mathcal{A}_{MX}\right)_{(8:10)}$$

and matrix $\mathcal{B}$ is

$$\mathcal{B} = C_{XX}^{-1}\left(\mathcal{G}_{MM}^{-1}\mathcal{A}_{MS}\right)_{(8:10)}$$

In these equations the subscript notation (10,11) denotes the 10-th and 11-th rows of the matrix. Since the output of interest is $v_O = v_4$, the output matrix $\mathcal{Y}$ is

$$\mathcal{Y} = v_O = v_4 = \mathcal{V}_M(4) = \left(\mathcal{G}_{MM}^{-1}\mathcal{A}_{MX}\right)_{(4)}X + \left(\mathcal{G}_{MM}^{-1}\mathcal{A}_{MS}\right)_{(4)}X_S = CX + \mathcal{D}X_S$$

Then

$$C = \left(\mathcal{G}_{MM}^{-1}\mathcal{A}_{MX}\right)_{(4)}$$

$$\mathcal{D} = \left(\mathcal{G}_{MM}^{-1}\mathcal{A}_{MS}\right)_{(4)}$$

```
% Example 4.5 - Third-Order Butterworth Filter Transient
%     File a - Computes and plots output voltage
clear
global xs A B rA
Ex_4_5b % Set up A and B
[ra,ca]=size(A);
clg, hold off
xs=1;
to=0; tf=10;
xo=[0 0 0]';
[t,x]=ode23('StVarEq',to,tf,xo);
plot(t,x(:,1))
xlabel('Time in ms')
ylabel('Output Voltage in V')
grid
hold
xc=1-exp(-t)-2.0.*exp(-t./2).*sin(sqrt(3).*t./2)/sqrt(3);
plot(t,xc,'*')
hold off
```

Figure 4.11 M–File to Plot Output Voltage vs. Time.

```
% Example 4.5 Third-Order Butterworth Transient
%      File b - Computes state-variable matrices A, B, C, and D
G1=1;  G2=1;  G3=1;
C1=1;  C2=1;  C3=1;
Gmm=nmAcc(1,5,G1,zeros(10));          Gmm=nmAcc(2,4,G2,Gmm);
Gmm=nmAcc(3,4,G3,Gmm);                Gmm=mnaVSrc(6,1,0,Gmm);
Gmm=mnaOA(7,0,2,3,Gmm);               Gmm=mnaVSrc(8,5,2,Gmm);
Gmm=mnaVSrc(9,5,3,Gmm);               Gmm=mnaVSrc(10,4,0,Gmm);
Cxx(1,1)=C1;        Cxx(2,2)=C2;      Cxx(3,3)=C3;
Amx([8:10],:)=eye(3);
Ams=zeros(10,1);                      Ams(6,1)=1;
Amx=Gmm\Amx;                          Ams=Gmm\Ams;
% The A, B, C, And D matrices are...
A=Cxx\Amx([8:10],:);                  B=Cxx\Ams([8:10],:);
C=[1 0 0];                            D=0;
```

Figure 4.12 M–File to Compute State-Variable Matrices $\mathcal{A}$ and $\mathcal{B}$ for Third-Order Butterworth Filter.

The M-file Ex_4_5a.M in Fig. 4.11, runs M-file Ex_4_5b.M in Fig. 4.12 to compute the state-variable matrices, then uses ode23 and the function StVarEq in Fig. 4.9 to compute and plot the output voltage. Figure 4.13 shows the resulting plot. The asterisks indicate the theoretical solution.

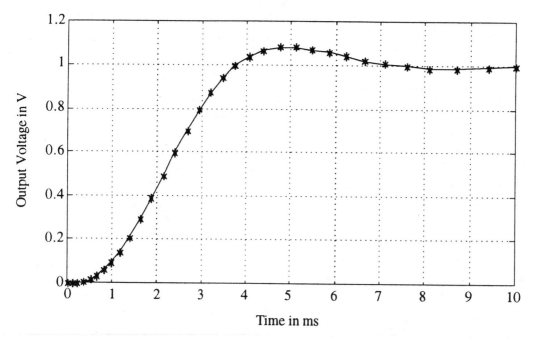

Figure 4.14 Transient Response of Third-Order Butterworth Filter to 1-V Step.

4.4 Residue   Calculation

Laplace transform solution of lumped, linear, time-invariant (LLTI) circuits finds solutions X(s) in the s domain, which have the form

$$X(s) = \frac{n_{nn}s^{nn} + n_{nn-1}s^{nn-1} + \ldots + n_1 s + n_0}{d_{nd}s^{nd} + d_{nd-1}s^{nd-1} + \ldots + d_1 s + d_0} = \frac{\sum_{j=0}^{j=nn} n_j s^j}{\sum_{j=0}^{j=nd} d_j s^j} \qquad (4.5)$$

The next step in the solution resolves the Laplace transform X(s) into the partial fraction expansion

$$X(s) = \frac{r_1}{s-p_1} + \frac{r_2}{s-p_2} + \ldots + \frac{r_{nd}}{s-p_{nd}} + k_{nn-nd}s^{nn-nd} + k_{nn-nd-1}s^{nn-nd-1} + \ldots + k_1 s + k_0$$

$$= \sum_{j=1}^{j=nd} \frac{r_j}{s-p_j} + \sum_{j=0}^{j=nn-nd} k_j s^j$$

$$(4.6)$$

where r_j is the residue of pole p_j and remainder coefficients k_k exist when the order of the numerator polynomial nn equals or exceeds the order of the denominator polynomial nd. Then, the solution x(t) is

$$x(t) = \mathcal{L}^{-1}\{X(s)\} = \sum_{j=1}^{nd} r_j e^{p_j t} u(t) + \sum_{j=0}^{nn-nd} k_j u_j(t) \qquad (4.7)$$

where u(t) is the unit-step function and $u_j(t)$ is the general singularity function. In particular, $u_0(t)$ is the unit-impulse function $\delta(t)$ and $u_1(t)$ is the unit doublet. The residue function

```
»[r,p,k]=residue([n_nn ... n_1 n_0],[d_nd ... d_1 d_0])
```

places the residues r_j in the array r, the poles p_j in array p, and remainder coefficients k_j in array k.

If a pole p_j has order k, then the partial fraction expansion contains the terms

$$X_j(s) = \frac{r_{j1}}{(s-p_j)^k} + \frac{r_{j2}}{(s-p_j)^{k-1}} + \ldots + \frac{r_{j1}}{s-p_j} = \sum_{q=1}^{k} \frac{r_{jq}}{(s-p_j)^q} \qquad (4.8)$$

The inverse Laplace transform of $X_j(s)$ is

$$x_j(t) = \sum_{q=1}^{k} \frac{r_{jq}}{(q-1)!} t^{q-1} \exp(p_k t) u(t) \qquad (4.9)$$

When a multiple-order pole exists, the residue function continues to return the residues of the single-order poles. However to determine the residues r_{jq} of the multiple-order poles, use

```
»rjq=residue(num,den,pj,k,q)
```

for values of q equal to 1, 2, ..., k to determine the k residues rjq.

Example 4.6 Impulse response of Third-Order Butterworth Circuit

For the third-order Butterworth circuit in Fig. 4.10, use the MATLAB residue function to compute the residues and poles of output-voltage Laplace transform $V_O(s)$ when the input voltage $v_S(t)$ is

   a) $\delta(t)$ V-ms

   b) $u(t)$ V

   c) $\exp(-t/2)\cos(\sqrt{3}\ t/2)u(t)$ V

For each set of residues and poles, write the solution for the output voltage.

Solution

First, determine the Laplace transform of the output voltage using nodal analysis at nodes 2, 4, and 5

$$
\begin{bmatrix}
0 & -G_2 & -sC_1 \\
-G_3 & sC_3 + G_2 + G_3 & -G_3 \\
-sC_1 & 0 & s(C_1 + C_2) + G_1
\end{bmatrix}
*
\begin{bmatrix}
V_{n3} \\
V_{n4} \\
V_{n5}
\end{bmatrix}
=
\begin{bmatrix}
0 \\
0 \\
G_1 V_s
\end{bmatrix}
$$

Solving for the output voltage $V_O = V_{n5}$ with $G_1 = G_2 = G_3 = G = 1$ mS and $C_1 = C_2 = C_3 = 1\mu F$ gives

$$
V_o(s) = \frac{\left(\dfrac{G}{C}\right)^3}{s^3 + 2\dfrac{G}{C}s^2 + 2\left(\dfrac{G}{C}\right)^2 s + \left(\dfrac{G}{C}\right)^3} \times V_s(s) = \frac{1}{s^3 + 2s^2 + 2s + 1} \times V_s(s)
$$

where the values use the units V, mA, ms, krad/s, kΩ, and μF.

a) When $v_S(t)$ is a unit-impulse function $\delta(t)$, then $V_S(s)$ equals 1 V-ms and

$$
V_o(s) = \frac{1}{s^3 + 2s^2 + 2s + 1}
$$

The numerator polynomial is num = [1] and the denominator polynomial is den = [1 2 2 1]. Using residue gives

```
»[r,p,k]=residue([1],[1 2 2 1])
r =
    1.0000
   -0.5000 - 0.2887i
   -0.5000 + 0.2887i
p =
   -1.0000
   -0.5000 + 0.8660i
   -0.5000 - 0.8660i
k =
    []
```

so the Laplace transform $V_O(s)$ is

$$V_o(s) = \frac{1}{s+1} - \frac{0.5000 + j0.2887}{s+0.5000 - j0.8660} - \frac{0.5000 - j0.2887}{s+0.5000 + j0.8660}$$

Since

$$0.5000 + j0.2887 = 0.5774\angle 30° = \frac{1}{\sqrt{3}}\angle 30°$$

the inverse Laplace transform gives

$$v_o(t) = \left\{ e^{-t} - \frac{2}{\sqrt{3}} e^{-t/2} \cos\left(\frac{\sqrt{3}}{2} t\right) \right\} u(t) \text{ V}$$

b) When $v_S(t)$ is a unit-step function $u(t)$, then $V_S(s)$ equals 1/s V-ms and

$$V_o(s) = \frac{1}{s^4 + 2s^3 + 2s^2 + s}$$

In this case, the numerator polynomial is num = [1] and the denominator polynomial is den = [1 2 2 1 0]. Using residue gives

```
»[r,p,k]=residue([1],[1 2 2 1 0])
r =
  -1.0000
   0.0000 + 0.5774i
   0.0000 - 0.5774i
   1.0000 + 0.0000i
p =
  -1.0000
  -0.5000 + 0.8660i
  -0.5000 - 0.8660i
        0
k =
      []
```

so the Laplace transform $V_O(s)$ is

$$V_o(s) = \frac{1}{s} - \frac{1}{s+1} + \frac{j0.5774}{s+0.5000 - j0.8660} - \frac{j0.5774}{s+0.5000 + j0.8660}$$

Taking the inverse Laplace transform gives

$$v_o(t) = \left\{ 1 - e^{-t} - \frac{2}{\sqrt{3}} e^{-t/2} \sin\left(\frac{\sqrt{3}}{2} t\right) \right\} u(t) \text{ V}$$

c) When $v_S(t)$ is $\exp(-t/2)\cos(\sqrt{3}\ t/2)u(t)$ V, the Laplace transform $V_S(s)$ is

$$V_s(s) = \frac{s + \dfrac{1}{2}}{\left(s + \dfrac{1}{2}\right)^2 + \left(\dfrac{\sqrt{3}}{2}\right)^2} = \frac{s + \dfrac{1}{2}}{s^2 + s + 1} \text{ V} - \text{ms}$$

Then the Laplace transform of the output voltage is

$$V_o(s) = \frac{1}{s^3 + 2s^2 + 2s + 1} \times \frac{s + \frac{1}{2}}{s^2 + s + 1} = \frac{s + \frac{1}{2}}{s^5 + 3s^4 + 5s^3 + 5s^2 + 3s + 1}$$

In this case, the numerator polynomial is num = [1 0.5] and the denominator polynomial is den = [1 3 5 5 3 1]. Using residue gives

```
»[r,p,k]=residue([1 .5],[1 3 5 5 3 1]);
»p
p =
  -0.5000 + 0.8660i
  -0.5000 - 0.8660i
  -1.0000
  -0.5000 + 0.8660i
  -0.5000 - 0.8660i
»r(3)
ans =
  -0.5000 + 0.0000i
```

MATLAB's response shows that the residue of the pole at -1 kNep/s is -0.5 V-ms. The pole listing shows that repeated roots of order 2 exist at -0.5±j0.8660 krad/s. To find r_{j1} and r_{j2} for the repeated root, use

```
»rj1=resi2([1 .5],[1 3 5 5 3 1],p(1),2,1)
rj1 =
   0.2500 - 0.1443i
»rj1=resi2([1 .5],[1 3 5 5 3 1],p(1),2,2)
rj1 =
  -0.2500 - 0.1443i
```

Then the Laplace transform of the output voltage is

$$V_o(s) = \frac{1}{s+1} - \frac{0.2500 + j0.1443}{(s + 0.5000 - j0.8660)^2} - \frac{0.2500 - j0.1443}{(s + 0.5000 + j0.8660)^2}$$
$$+ \frac{0.2500 - j0.1443}{s + 0.5000 - j0.8660} + \frac{0.2500 + j0.1443}{s + 0.5000 + j0.8660}$$

Since

$$0.2500 \pm j0.1443 = 0.2887 \angle \pm 30° = \frac{1}{2}\frac{1}{\sqrt{3}} \angle \pm 30°$$

then taking the inverse Laplace transform gives

$$v_o(t) = \left\{ -\frac{1}{2}e^{-t} - \frac{1}{\sqrt{3}}e^{-t/2}\left[t\cos\left(\frac{\sqrt{3}}{2}t + 30°\right) - \cos\left(\frac{\sqrt{3}}{2}t - 30°\right)\right] \right\} u(t) \text{ V}$$

You can easily check MATLAB's solution for Part a) and Part b) by finding the residues using the formula

$$r_k = (s - p_k)V_o(s)\big|_{s=p_k}$$

for each pole p_k. To check the residues r_{kq} for the repeated root in Part c), you can use

$$r_{kq} = \frac{1}{(q-1)!} \frac{d^{q-1}}{ds^{q-1}} \left\{ (s - p_k)^k V_o(s) \right\} \Big|_{s=p_k}$$

which leads to a substantial amount of work.

Appendix A

CK_TOOL Functions

```
function ncols = cols(aMat)
%cols(aMat) returns the number of columns in matrix aMat.
[nrows,ncols] = size(aMat);

function nrows = rows(aMat)
%rows(aMat) returns the number of rows in matrix aMat.
[nrows,ncols] = size(aMat);

function xx = genAnal(xs,axx,axd,axs,adc,acd,acx,acs)
% genAnal(xs,axx,axd,axs,adc,acd,acx,acs)
%   Calculates the node-voltage or mesh-current vector
%   xs  - Source vector
%   axx - Node or mesh matrix
%   axd - Dependent source matrix
%   adc - Relates dependent sources to controlling variables
%   acd - Relates controlling variables to dependent sources
%   acx - Relates controlling variables to mesh voltages or
%         mesh currents
%   acs - Relates controlling variables to independent sources
[m,n] = size(adc);
b = axd*((eye(m)-adc*acd)\adc);
xx = (axx+b*acx)\(axs-b*acs)*xs;

function amat = gmAcc(j,jp,k,kp,val,bmat)
%gmAcc(j,jp,k,kp,val,bmat)
%    Accumulates gm value "val" onto matrix bmat.
%    j and jp are + and - control nodes.
%    k and kp are from and to nodes.
if (j<0)|(jp<0)|(k<0)|(kp<0)
     disp('Indices cannot be negative')
elseif ((j==0)&(jp==0))|((k==0)&(kp==0))
     disp('A trivial circuit element')
else
     if (k>0)&(j>0)
          bmat(k,j) = bmat(k,j)+val;
     end
     if (kp>0)&(jp>0)
          bmat(kp,jp) = bmat(kp,jp)+val;
     end
     if (k>0)&(jp>0)
          bmat(k,jp) = bmat(k,jp)-val;
     end
     if (kp>0)&(j>0)
          bmat(kp,j) = bmat(kp,j)-val;
     end
end
amat = bmat;
```

```
function amat = mnaESrc(m,j,jp,k,kp,Eval,bmat)
%mnaESrc (m,j,jp,k,kp,Eval,bmat)
%  Accumulation of Esrc having value "Eval" onto bmat.
%  m is the index of the added CE row.
%  j and jp are the + and - controlling nodes.
%  k and kp are the + and - nodes of the E source.
if ((m<0)|(j<0)|(jp<0)|(k<0)|(kp<0))
     disp('Indices cannot be negative')
else
     if (k>0)
          bmat(k,m) = 1;
          bmat(m,k) = 1;
     end
     if (kp>0)
          bmat(kp,m) = -1;
          bmat(m,kp) = -1;
     end
     if (j>0)
          bmat(m,j) = bmat(m,j)-Eval;
     end
     if (jp>0)
          bmat(m,jp) = bmat(m,jp)+Eval;
     end
end
amat = bmat;


function amat = mnaFSrc(m,j,jp,k,kp,Fval,bmat)
%mnaFSrc (m,j,jp,k,kp,Fval,bmat)
%  Accumulation of Fsrc value "Fval" onto bmat.
%  m is the index of the added CE row.
%  j and jp are the from and to controlling nodes.
%  k and kp are the from and to nodes of the F source.
if ((m<0)|(j<0)|(jp<0)|(k<0)|(kp<0))
     disp('Indices cannot be negative')
else
     if (j>0)
          bmat(j,m) = 1;
          bmat(m,j) = 1;
     end
     if (jp>0)
          bmat(jp,m) = -1;
          bmat(m,jp) = -1;
     end
     if (k>0)
          bmat(k,m) = bmat(k,m)+Fval;
     end
     if (kp>0)
          bmat(kp,m) = bmat(kp,m)-Fval;
     end
end
amat = bmat;
```

```
function amat = mnaHSrc(m,j,jp,k,kp,Hval,bmat)
%mnaHSrc(m,j,jp,k,kp,Hval,bmat)
%  Accumulation of Hsrc value "Hval" onto bmat.
%  m is the index of the added CE row.
%  j and jp are the from and to controlling nodes.
%  k and kp are the + and - nodes of the H source.
if ((m<0)|(j<0)|(jp<0)|(k<0)|(kp<0))
      disp('Indices cannot be negative')
else
      if j>0,
            bmat(j,m) = 1;
            bmat(m,j) = 1;
      end
      if jp>0,
            bmat(jp,m) = -1;
            bmat(m,jp) = -1;
      end
      if k>0,
            bmat(k,m+1) = 1;
            bmat(m+1,k) = 1;
      end
      if kp>0,
            bmat(kp,m+1) = -1;
            bmat(m+1,kp) = -1;
      end
      bmat(m+1,m) = -Hval;
end
amat = bmat;


function amat = mnaIdX(m,j,jp,k,kp,nVal,bmat)
%mnaIdX(m,j,jp,k,kp,nVal,bmat)
%  Accumulation of ideal transformer with turns ratio nVal
%  onto bmat. This function adds two rows to your matrix.
%     m is the index of the first additional CE row.
%     j (dot) and jp are the from and to input branch nodes.
%     k (dot) and kp are the from and to output branch nodes.
if ((m<0)|(j<0)|(jp<0)|(k<0)|(kp<0))
      disp('Indices cannot be negative')
else
      if j>0,
            bmat(j,m) = 1;
            bmat(m+1,j) = -nVal;
      end
      if jp>0,
            bmat(jp,m) = -1;
            bmat(m+1,jp) = -nVal;
      end
      if k>0,
            bmat(k,m+1) = 1;
            bmat(m+1,k) = bmat(m+1,k)+1;
      end
```

```
        if kp>0,
                bmat(kp,m+1) = -1;
                bmat(m+1,kp) = bmat(m+1,kp)-1;
        end
        bmat(m,m) = 1;
        bmat(m,m+1) = nVal;
end
amat = bmat;


function amat = mnaIdX1(m,j,jp,k,kp,nVal,bmat)
%mnaIdX1(m,j,jp,k,kp,nVal,bmat)
%  Accumulation of ideal transformer with turns ratio nVal
%  onto bmat. This function adds only one additional row
%  to your matrix.
%     m is the index of the one additional row.
%     j (dot) and jp are the from and to input branch nodes.
%     k (dot) and kp are the from and to output branch nodes.
if ((m<0)|(j<0)|(jp<0)|(k<0)|(kp<0))
        disp('Indices cannot be negative')
else
        if j>0,
                bmat(j,m) = -nVal;
                bmat(m,j) = -nVal;
        end
        if jp>0,
                bmat(jp,m) = nVal;
                bmat(m,jp) = nVal;
        end
        if k>0,
                bmat(k,m) = bmat(k,m)+1;
                bmat(m,k) = bmat(m,k)+1;
        end
        if kp>0,
                bmat(kp,m) = bmat(kp,m)-1;
                bmat(m,kp) = bmat(m,kp)-1;
        end
end
amat = bmat;


function amat = mnaLM(m,L1,L2,M,bmat)
%mnaLM(m,L1,L2,M,bmat)
%  Accumulation of L1, L2, and M onto bmat
%  m is the index of the first of two additional CE rows.
if (m<0)
        disp('Index cannot be negative')
else
        bmat(m,m) = -L1;
        bmat(m,m+1) = -M;
        bmat(m+1,m) = -M;
        bmat(m+1,m+1) = -L2;
end
amat = bmat;
```

```
function amat = mnaOA(m,j,jp,k,bmat)
%mnaOA(m,j,jp,k,bmat)
%   Accumulation of entries for an ideal op amp onto bmat.
%       m is the index into the added row.
%       j and jp are the noninverting and the inverting op-amp
%         input nodes.
%       k is the output node.
if ((m<0)|(j<0)|(jp<0)|(k<0))
    disp('Indices cannot be negative')
else
    if j>0,
        bmat(m,j) = 1;
    end
    if jp>0,
        bmat(m,jp) = -1;
    end
    if k>0,
        bmat(k,m) = 1;
    end
end
amat = bmat;



function amat = mnaVSrc(m,i,j,bmat)
%mnaVSrc(m,i,j,bmat)
% Accumulation of Vsrc onto bmat for MNA analysis.
%    m is the index into the additional CE row.
%    i and j are the + and - nodes of the voltage source.
if ((i<0)|(j<0)|(m<0))
    disp('Indices cannot be negative')
else
    if (i>0)
        bmat(m,i) = 1;
        bmat(i,m) = 1;
    end
    if (j>0)
        bmat(m,j) = -1;
        bmat(j,m) = -1;
    end
end
amat = bmat;



function amat = nmAcc(i,j,val,bmat)
%nmAcc(i,j,val,bmat)
% Accumulates G or R values "val" onto bmat.
% i and j are node (mesh) indices.
if (i<0)|(j<0)
    disp('row/col indices cannot be negative')
elseif  (i==0)&(j==0)
    disp('Both indices equal 0. Trivial!')
```

```
    else
        if (i>0)
            bmat(i,i) = bmat(i,i)+val;
        end
        if (j>0)
            bmat(j,j) = bmat(j,j)+val;
        end
        if (i>0) & (j>0)
            bmat(i,j) = bmat(i,j)-val;
            bmat(j,i) = bmat(j,i)-val;
        end
end
amat = bmat;


function amat = srcAcc(i,j,val,bmat)
%srcAcc(i,j,val,bmat)
%    Accumulates Is or Vs values onto source matrix bmat.
%       Node i is "from" node (drop mesh).
%       Node j is "to" node (rise mesh).
if (i<0)|(j<0)
    disp('From/To indices cannot be negative')
elseif (i==0)&(j==0)
    disp('Both indices equal zero. Trivial!')
else
        if (j>0)
            bmat(j) = bmat(j)+val;
        end
        if (i>0)
            bmat(i) = bmat(i)-val;
        end
end
amat = bmat;


function xdot = StVarEq(t,x)
%StVarEq(t,x)
%  Define source vector xs, state variable matrices A and B,
%  and rows rA of A as global variables.
%  Comment out the statement "global xs A B rA" if using
%  The Student Edition of MATLAB 3.5, but define xs, A, B,
%  and rA in your workspace before calling StVarEq.
global xs A B rA
xdot=zeros(rA,1);
xdot=A*x+B*xs;
```

Index

YOU SHOULD CAREFULLY READ THE FOLLOWING TERMS AND CONDITIONS BEFORE OPENING THIS DISKETTE PACKAGE. OPENING THIS DISKETTE PACKAGE INDICATES YOUR ACCEPTANCE OF THESE TERMS AND CONDITIONS. IF YOU DO NOT AGREE WITH THEM, YOU SHOULD PROMPTLY RETURN THE PACKAGE UNOPENED, AND YOUR MONEY WILL BE REFUNDED.

IT IS A VIOLATION OF COPYRIGHT LAWS TO MAKE A COPY OF THE ACCOMPANYING SOFTWARE EXCEPT FOR BACKUP PURPOSES TO GUARD AGAINST ACCIDENTAL LOSS OR DAMAGE.

Prentice-Hall, Inc. provides this program and licenses its use. You assume responsibility for the selection of the program to achieve your intended results, and for the installation, use, and results obtained from the program. This license extends only to use of the program in the United States or countries in which the program is marketed by duly authorized distributors.

LICENSE

You may:

a.   use the program;
b.   modify the program and/or merge it into another program in support of your use of the program.

LIMITED WARRANTY

THE PROGRAM IS PROVIDED "AS IS" WITHOUT WARRANTY OF ANY KIND, EITHER EXPRESSED OR IMPLIED, INCLUDING, BUT NOT LIMITED TO, THE IMPLIED WARRANTIES OF MERCHANTABILITY AND FITNESS FOR A PARTICULAR PURPOSE. THE ENTIRE RISK TO THE QUALITY AND PERFORMANCE OF THE PROGRAM IS WITH YOU. SHOULD THE PROGRAM PROVE DEFECTIVE, YOU (AND NOT PRENTICE-HALL, INC. OR ANY AUTHORIZED DISTRIBUTOR) ASSUME THE ENTIRE COST OF ALL NECESSARY SERVICING, REPAIR, OR CORRECTION.

SOME STATES DO NOT ALLOW THE EXCLUSION OF IMPLIED WARRANTIES, SO THE ABOVE EXCLUSION MAY NOT APPLY TO YOU. THIS WARRANTY GIVES YOU SPECIFIC LEGAL RIGHTS AND YOU MAY ALSO HAVE OTHER RIGHTS THAT VARY FROM STATE TO STATE.

Prentice-Hall, Inc. does not warrant that the functions contained in the program will meet your requirements or that the operation of the program will be uninterrupted or error free.

However, Prentice-Hall, Inc., warrants the diskette(s) on which the program is furnished to be free from defects in materials and workmanship under normal use for a period of ninety (90) days from the date of delivery to you as evidenced by a copy of your receipt.

LIMITATIONS OF REMEDIES

Prentice-Hall's entire liability and your exclusive remedy shall be:

1.   the replacement of any diskette not meeting Prentice-Hall's "Limited Warranty" and that is returned to Prentice-Hall with a copy of your purchase order, or

2.   if Prentice-Hall is unable to deliver a replacement diskette or cassette that is free of defects in materials or workmanship, you may terminate this Agreement by returning the program, and your money will be refunded.

IN NO EVENT WILL PRENTICE-HALL BE LIABLE TO YOU FOR ANY DAMAGES, INCLUDING ANY LOST PROFITS, LOST SAVINGS, OR OTHER INCIDENTAL OR CONSEQUENTIAL DAMAGES ARISING OUT OF THE USE OR INABILITY TO USE SUCH PROGRAM EVEN IF PRENTICE-HALL, OR AN AUTHORIZED DISTRIBUTOR HAS BEEN ADVISED OF THE POSSIBILITY OF SUCH DAMAGES, OR FOR ANY CLAIM BY ANY OTHER PARTY.

SOME STATES DO NOT ALLOW THE LIMITATION OR EXCLUSION OF LIABILITY FOR INCIDENTAL OR CONSEQUENTIAL DAMAGES, SO THE ABOVE LIMITATION OR EXCLUSION MAY NOT APPLY TO YOU.

GENERAL

You may not sublicense, assign, or transfer the license or the program except as expressly provided in this Agreement. Any attempt otherwise to sublicense, assign, or transfer any of the rights, duties, or obligations hereunder is void.

This Agreement will be governed by the laws of the State of New York.

Should you have any questions concerning this Agreement, you may contact Prentice-Hall, Inc., by writing to:

Prentice Hall
College Division
Englewood Cliffs, NJ 07632

YOU ACKNOWLEDGE THAT YOU HAVE READ THIS AGREEMENT, UNDERSTAND IT, AND AGREE TO BE BOUND BY ITS TERMS AND CONDITIONS. YOU FURTHER AGREE THAT IT IS THE COMPLETE AND EXCLUSIVE STATEMENT OF THE AGREEMENT BETWEEN US THAT SUPERSEDES ANY PROPOSAL OR PRIOR AGREEMENT, ORAL OR WRITTEN, AND ANY OTHER COMMUNICATIONS BETWEEN US RELATING TO THE SUBJECT MATTER OF THIS AGREEMENT.

ISBN: 0-13-127044-3

We hope **Matrix Analysis of Circuits Using MATLAB**®· by James Gottling will meet your classroom needs. Please take a moment to complete the following information, then mail the card to us so we may learn more about your department. Thank you!

Name _____ Phone _____

School Address _____

Department _____

Office Hours _____

Course Title and Number _____

Current Text _____

Length of Course _____

Enrollment: _____ Fall _____ Spring _____ Other _____

Are you likely to change books? _____ Decision Date _____

Other decision makers? _____

Are you currently using software? _____ If yes, what kind? _____

Do you want to see information regarding **Linear Circuit Analysis: Time Domain, Phasor, and Laplace Transform Approaches** by DeCarlo and Lin, which this Manual supplements? _____

Comments _____

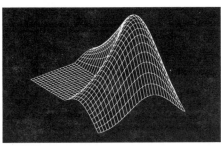

NAME _____

TITLE _____

COMPANY _____

DEPT. OR M/S _____

STREET _____

CITY/STATE/ZIP _____

PHONE _____

FAX _____ EMAIL _____

WHERE DID YOU PURCHASE THIS BOOK? _____

Computer platform – check all that apply:

❑ PC/Macintosh   ❑ UNIX Workstation   ❑ VAX/Supercomputer

▶ **For the fastest response, fax this card to:**
(508) 653-6284, or call us at (508) 653-1415.

MATLAB® Technical Computing Environment

R-BK-GOT

MATLAB, the companion software to **Matrix Analysis of Circuits Using MATLAB** by James G. Gottling (Prentice Hall, 1995), is used in this book's presentation of the theory, and for problem-solving. Widely used in academia, industry, and government, MATLAB has become the premier technical computing environment for electrical engineering, applied math, physics, and other disciplines.

- **MATLAB Application Toolboxes** add functions for symbolic math, signal processing, control design, neural networks, and other areas.

- **The Student Edition** is a limited-matrix-size version of MATLAB for use on students' own personal computers.

- **Educational discount plans** support classroom instruction and research.

- **Classroom Kits** provide cost-effective support for PC or Mac teaching labs.

- **MATLAB-based books** use MATLAB to illustrate basic and advanced material in a wide range of topics.

I am interested in The MathWorks product information for:

❑ Simulation   ❑ Control System Design   ❑ Math & Visualization
❑ Signal Processing ❑ Symbolic Math   ❑ Educational Discounts
❑ System Identification ❑ Chemometrics   ❑ Classroom Kits
❑ Neural Networks   ❑ Optimization   ❑ Student Edition
❑ Statistics   ❑ Image Processing   ❑ MATLAB Books

PLACE
STAMP
HERE

Marketing Manager
Engineering/Computer Science
PRENTICE HALL
Simon & Schuster Education Group
113 Sylvan Avenue, Route 9W
Englewood Cliffs, NJ 07632

The
MATH
WORKS
Inc.

NO POSTAGE
NECESSARY IF
MAILED IN THE
UNITED STATES

BUSINESS REPLY MAIL

FIRST CLASS PERMIT NO. 82 NATICK, MA

POSTAGE WILL BE PAID BY ADDRESSEE

The MathWorks, Inc.
24 Prime Park Way
Natick, MA 01760-9889